Interviews with Key Figures of 20 Years' Xiamen ICM
1996 – 2016

Edited by Wang Chunsheng
Nian Yue

By Xiamen Municipal Bureau of Oceans and Fisheries
     Taihai Magazine

Wake up the ears
listen to the story of the Xiamen sea

CONTENTS

001 Coastal Management Should Revere Nature
Zhu Yayan

020 "Revering the Sea" in the Context of Urban Construction
Pan Shijian

045 The Xiamen Model Is of Epoch-Making Significance in the History of World Coastal Management
Chua Thia-Eng

Interviews with Key Figures of 20 Years Xiamen ICM
1996-2016

061   Export Integrated Coastal Management Experience from Xiamen to the World
Wang Chunsheng

104   Collaboration Is the Core Idea of Integrated Coastal Zone Management
Ruan Wuqi

121   Coordination Between Agencies Is Essential for Marine Management
Lin Hanzong

132   Strong Science and Technology Bases for Integrated Management
Hong Huasheng

146   Marine Expert Panel Provides Robust Technical Support
Xue Xiongzhi

161   Xiamen Is the Birthplace of Marine Research of the Chinese Mainland
Li Shaojing

CONTENTS

175	Coordination Between Marine and Land Administration to Propel Ecological Rehabilitation
	Yu Xingguang

193	Multi-pronged Methods to Restore Marine Ecology
	Huang Chaoqun

205	The Success of Yundang Lake Case Lies in Integrated Management
	Zhang Bin

227	Management of lake Is to Protect the Lung of Xiamen
	Lin Xueping

239	The Economy of Xiamen features the Marine Economy
	Zheng Jinmu

251	Development of All Forms Should Be Based on Protection
	Guo Yunmou

265	Promoting the Sustainable Development of Marine Economy
	Peng Benrong

280	Riding the Wind and Waves and Piloting Marine Management — Experience from Integrated Coastal Zone Management of Xiamen

294	Summary of Scientific Research Programs During 9th Five-Year Plan Period to 12th Five-Year Plan Period of Marine and Island management Office, Xiamen Municipal Bureau of Oceans and Fisheries

Coastal Management Should Revere Nature

Zhu Yayan

In 1986, Hui'an-born Zhu Yayan, at the time working in an institution affiliated with the Ministry of Chemical Industry, applied to return to Southern Fujian to accompany his elderly mother. Both Quanzhou and Xiamen responded. Zhu asked the Ministry which city could accommodate him faster. The answer was "Xiamen. " Zhu responded readily, "Then I will go to Xiamen. "Before long, Zhu Yayan returned to Xiamen.

Next year, Zhu was elected Vice Mayor of Xiamen city, and then went on to serve as the Executive Vice Mayor, Mayor, Vice Governor of Fujian Province, member of the Provincial Party Committee, Secretary of the Provincial Party Committee, Deputy Director of the Standing Committee of Provincial People's Congress, etc. He worked in Xiamen for over 10 years, and stayed there after retirement.

Hui'an brings to mind the vision of the sea, fishing boats and hard-working Hui'an women. Zhu Yayan said frankly, when he was a teenager and saw the sea for the first time, he was not excited at all; so when he was assigned to Sichuan province by the Ministry of Chemical Industry and was eager to come back, it was not because he yearned to see the ocean again. However, it is this man who claimed to have no special feeling for the ocean that pioneered the coastal management of Xiamen. Before Zhu Yayan, no one in Xiamen or other coastal cities really had a clear understanding of coastal management. When Zhu recalled the year when they put forward the motion of coastal management, he said, "Only experts realized at the beginning how important coastal zone management is, how much benefit coastal management will bring to city development and how coastal management is an integral part of the blueprint for the development of marine economy. As no experts, we failed to see the significance at that time." He said, in the beginning, he believed coastal management was necessary only because the marine order was chaotic, and to do nothing would be unacceptable.

It is in a large part due to this simple idea that Xiamen has such a beautiful coast today, and the system of coastal management brings inestimable benefits to the city. At the same time, the Integrated Coastal Management of Xiamen has also drawn worldwide attention and become a model of the UN.

Zhu Yayan in his seventies now seldom goes to the seaside. Living beside the Yundang Lake, he feels at ease when the sea breeze carries over the pleasant aromas of clean ocean air, and worried when it has a foul smell.

In the summer of 2016, as the sun shone brightly and the wind blew gently, this old man, who devoted his working life to the ocean, recalled the past beside Yundang Lake.

"Revere nature—this is an important lesson that the ocean has taught me. " Zhu exhorted.

Introduction to the speaker

Zhu Yayan served as Executive Deputy Mayor and Mayor of Xiamen, Vice Governor of Fujian Province and deputy director of the Standing Committee of the Povincial People's Congress. During his career in Xiamen, he was involved in marine management. He once worked as the chief of Xiamen Integrated Coastal Management Leading Group, and chaired Xiamen Ocean Governance Initiative. On April 23, 2002 he hosted the mobilization meeting for comprehensive management of aquaculture in the West Seas, which is a milestone for Xiamen Integrated Coastal Management.

▼ The silhouettes of egrets against the early evening sky
(photo / Yang Zhijun)

※ Making Preparations for Coastal Management after Learning a Lesson

Although I was born in Hui'an, the place that linked me inextricably to the ocean is Xiamen.

In the second half of 1987, I served as Vice Mayor of the Xiamen Municipal Government, in charge of traffic and electricity. From then on, I had a difficult time in dealing with the ocean of Xiamen, quite a difficult time.

What do I mean by "difficult time"? In those years, at the channels of Xiamen port, there

1
---
2   A comparison between the new and the old Xiagu

were always fishing boats working in a way that affected the flow of traffic. Fishermen thought of a clever way to catch eel: they first bought an old hen or a rooster and put it in a net after slaughtering it. Then they would fix the net to two boats, which were steered to the main channels of Xiamen port, whereby the fishermen tossed the net into the sea. Soon, young eels were attracted to the chicken bait and could be caught easily by the fishermen. Those years, one young eel could be sold for one yuan. This was an easier way to make money than catching fish. However, the ten-meter-long nets that hung across the channel caused severe traffic congestion. By the year 1988, the situation worsened. Some fishermen accidentally damaged the electrical cables laid at the bottom of the channel, leading to power cuts throughout Gulangyu Island. Complaints from locals were heard everywhere. Although air conditioners were not common at the time, most people used refrigerators to store vegetables since they tended to buy a lot of them at a time due to inconveniences. When the power went out, food easily spoiled, seriously affecting the daily lives of the citizens in Gulangyu Island.

These problems really bothered me, so I felt deeply the urgent need for marine order to be restored.

In the 1990s, in addition to the unorderly marine traffic, the ecological environment was deteriorating. The aquaculture sector wished to develop marine aquaculture, and promulgated the slogan "Creating a brand new Xiamen". As a result, the whole of Xiamen, including Dongdu Harbor and Huoshao Island, devoted itself to breeding shrimp. On the other hand, the tourism sector wished to develop marine tourism, so many ships opened restaurants along the coast, discharging sewage "naturally" into the marine environment. In addition, Xiamen in the 1990s had entered the phase of large-scale urban development. The problem of water and soil erosion worsened and the whole of Xiamen Harbor was burdened by sludge.

The marine problems of Xiamen were so bad that they affected my duties in entertaining and receiving important guests.

In the beginning of the 1990s, every time I received guests, I would first consult the lunar calendar in order to estimate tidal patterns. When we were about to encounter the ebb tide, I would try my best to change the visiting hour to a time when the tides were higher, because the shore during ebb tide was terribly ugly. Xiamen was hailed as a "marine garden". A common saying goes, "The sea is in the city and the city is on the sea". Our guests came here to see the beautiful scenery of Xiamen. Did we want to show them the sludge? Of course not.

We had problems with marine traffic and the ecological environment, but these problems were not fatal. The fatal problem was: everyone claimed authority, but no one took responsibility. As a result, marine problems accumulated and gradually worsened. Only experts realized at the beginning that coastal management was important and could bring benefits to city development as well as the development of the marine economy. As no experts, we failed to see the importance at that time. For civil servants in the municipal government, the original intension of coastal management was to manage the traffic in Xiamen Harbor. The reasons are as follows: first, Xiamen is a port, so marine commerce is the lifeblood of the city. Second, Xiamen is positioned as a tourist destination. If the marine environment were a mess and unbearable even to locals, it no longer deserved the name "marine garden". Third, coastal management should be sustainable. It is not something that when leaders mention it, is done for a short time, and when the leaders ignore it, is quickly forgotten.

This was the initial phase when officials of the Xiamen government developed a

▲ A picturesque view of the West Seas of Xiamen after rehabilitation

preliminary understanding of coastal management.

※ Setting up the Coastal Management Scheme after 20 Years of Experimentation

Throughout the 20 years of coastal management, we have made many accomplishments and provided some successful models for others. We have also been widely recognized by marine organizations at home and abroad. Doctor Chua Thia-Eng, former director of the PEMSEA once gifted me a hand-crafted memento after we shared our experience with them. The handicraft was a mother carrying her kid surrounded by green hills and blue water, expressing the theme of environmental protection. The State Oceanic Administration speaks highly of our work and selects Xiamen as the host city of the International Ocean Week each year.

What are the unique features of the coastal management of Xiamen? For me as a practitioner, the main features are as follows:

First, we began by discussing practical concerns, instead of holding meaningless meetings. A guiding principle and practical measures are both needed to determine what and how to manage. These are the primary issues. As leaders, we should not only have empty slogans like "raising awareness, unifying thoughts and strengthening leadership", but be specific

1. City leader conducted field research.
2. The first problem faced by the rehabilitation of the West Seas is the withdrawal of aquafarming to make way for a port of the Mingda Glass Plant.
3. Aquafarming cages lined the port and clogged the waterways.

as to what thoughts to be unified, what kind of leadership to be strengthened, and how to manage.

At first, there were two primary issues that constantly came up. One was *Xiamen Coastal Functional Zoning*. The coastline of Xiamen being over 200 kilometers long, we needed to figure out how to utilize it. The western sea area was damaged by aquaculture, an activity now forbidden. So we needed to launch new policies for fishermen to make an alternative living. The other issue was *Provisions on the Use of Sea Areas of Xiamen*. For example, if a person wants to use the sea or tidal-flat area, what procedures should he follow? How much should he pay for the use? How much should a user of the state construction pay, and how much should a user of aquaculture pay? I checked my working diary and found that we had finished the fifth draft by August to September in 1996. When these two regulations were approved during the executive meeting of the municipal government on 22nd October, they were at least the seventh or eighth draft.

Through the drafting process of these two regulations, we got a better understanding of the working methods and cut-in points of coastal management.

Additionally, in order to help unify the understanding of coastal management in all circles, I would invite newspapers and TV stations to cover our meetings, including Leading Group meetings and Standing Committee meetings. This management strategy would cover the work involving the Land Bureau, Marine Management Office, Planning Bureau, Urban Management Office, Tourist Administration, Harbor Bureau, Oceans and Fisheries Bureau, etc. So our first endeavor was launching specific and practical solutions, which clarified our management scope and unified the interests and considerations of all circles. Additionally, we took a step-by-step approach. In the course of implementation, we adopted some practical policies, with prompt policy subsidies and reasonable transitional periods. Therefore, citizens were more likely to accept our policies.

1 | 2
  | 3

1. On April 23, 2002, the mobilization meeting for withdrawal of aquaculture in the West Seas of Xiamen was held. This is a key turning point in the rehabilitation of Xiamen's West Seas.
2. Signing the responsibility guarantee
3. The swearing convention required complete withdrawal of aquafarming in the West Seas before October 31, 2002.

On April 23rd, 2002, before the Mobilization Meeting on the Prohibition of Aquaculture in the Western Sea Area, I was worried that some fishermen would petition against the policy, thus hindering the work of the provincial government. I went to Fuzhou city to communicate with provincial government officials. But afterwards, when I asked about any formal complaints, I was told that no fisherman had petitioned, a sign of our policy being relatively practical, thus winning people's support.

Second, we organized a coastal zone training course at Xiamen University and invited Qu Geping, "the pioneer of environmental protection in China", to give lectures. Mr. Qu didn't major in Environmental Studies in university. He graduated from the Department of Chinese Language and Literature of Shandong University and then went to Jilin University for further studies. In his research, he studied China's environmental protection issues and later he became the first director of the Ministry of Environmental

Protection of the People's Republic of China. Even though Xiamen is not large as coastal cities go, it is still part of his concern. He didn't hesitate to accept our invitation. Director Qu's arrival was a great inspiration; we all were very excited.

Our training course was held for several consecutive periods with a large number of trainees and high-level teachers, and functioned greatly to promote coastal management. In addition to inviting Qu Geping to give lectures, we also attached special importance to utilizing the experience of local marine experts for coastal management. We established a marine expert panel, covering a variety of subjects. The experts, coming from different institutions, differ in research focus and see things from different perspectives. Thus, they often had disputes and disagreements. Whenever disputes arose, I would never make decisions during the meeting, but wait until the meetings ended. Then I would talk with them separately. Everyone was well educated, but their biggest weakness was a desire to save face. We need to realize this when we do our job. Sometimes we would adopt non-mainstream opinions as long as they were reasonable.

1. Cleaning up silt in the Jimei sea (Photo / Li Wanhan)
2. The Jimei sea was silted up before rehabilitation. (Photo / Li Wanhan)

Third, another of Xiamen's experiences worthy of replicating is that while our different departments shoulder their respective responsibilities, we set up a special professional coastal management team and law enforcement team. In order to set up the Marine Management Office and professional law enforcement team, we spent a whole month recruiting staff and setting manning quotas. These teams represent the interests of different departments, each one having its own team. But what should we manage? The Marine Management Office and marine law enforcement team could work together. I checked my working diary. In August 1996, we discussed plans and measures of coastal management. In September, we set up the Marine Management Office and law enforcement team. In November, we settled the issues of legislation. Practice proves that it was effective to unify all forces together by setting up the Office and law enforcement team.

In fact, the Marine Management Office acts as an initiating unit, which lays the foundation work for coastal management and does not affect the motivations of other departments and units. Departments with more enthusiasm could be influenced a lot by the Management Office's directives and initiatives, whereas those with less enthusiasm could be influenced a little. With this solid foundation, we became more confident. In retrospect, at the beginning of the establishment of the Office, many comrades, including Yang Benxi and Wang Chunsheng, endured many hardships, but none of them gave up. Later, the Marine Management Office merged with the Aquatic Products Bureau, with their functions combined, into Xiamen Oceans and Fisheries Bureau the present pillar in coastal management.

Fourth, our sound laws, regulations and systems deserve to be mentioned. According to coastal management regulations and sea area planning frameworks, we set up a series of regulations and laws—the Harbor Bureau set up *Coastline Management Regulations and Professional Wharf Management Regulations*; Marine Affairs and the Surveillance Department set up *Marine Security Management Regulations*; the Environmental Protection Department set up *Provisions on Marine Environmental Protection of Xiamen*; the Land Department and Tourism Department also set up their respective regulations. Therefore, Doctor Chua Thia-Eng from PEMSEA showed great interest in our regulations and laws. He has been in this field for many years. Therefore, he clearly knows that no achievement could be made without practice, because coastal management is related to many departments and units. Once, PEMSEA held a meeting in Thailand. They invited us to share our experiences. I, along with some cadres, went to the meeting and delivered a keynote speech.

When we began issuing coastal management legislation, there was no set model for us to follow, so we mainly experimented on our own. Because government regulations are relatively easier to implement and change, many of the changes did not start from laws, but from rules and regulations. Today we believe that those regulations are outdated and insufficient, but at that time, they were the legal basis that ensured the implementation of Marine Functional Zoning and coastal management.

The last experience is the great attention we received from leaders. First of all, the State Oceanic Administration paid great attention to our work. Here I want to thank Doctor Chua Thia-Eng again for his support, both in the State Oceanic Administration and in PEMSEA. The Xiamen municipal government also laid much stress on it. They set up a leading group and appointed me as the group leader with at least three deputy mayors as vice group leaders. We held at least 3 meetings per month. The work was pushed forward intensively. In the government executive meeting on October 22nd, 1996, the coastal management proposal was passed unanimously thanks to our thorough preparations. In 1997, we put forward an objective: by the year 1998, the sea should be in good order.

Later, the government made some adjustments to this objective, because the leadership changed in 1997 and fishermen needed to be given some time for the closure of aquaculture. By the end of 1997, I was transferred to the provincial government, and at the beginning of 2000, I returned to Xiamen. In 2002, the coastal management of the western sea area started. The actual implementation did not progress as expected, mainly due to the change in leadership. In addition, owing to the implementation of other policies, progress slowed a bit during that period of transition, but our workload was not reduced.

※ Laying a Solid Foundation for Urban Development

It has been twenty years since we started work on coastal management. Throughout these years, we have experienced both gains and losses. Twenty years ago, when I first took over this job, I felt the work came a little bit too late. Now I realize that it was not late at all.

First, let's talk about what we have gained. Things have causes and effects. It is because of our work in coastal management that we were able to set the goal of building Xiamen into a port city in the municipal Party Congress on July 1st, 2000. The bay of Xiamen could not have been what it is like today if we had not divided the functions of the sea: the western sea area as an aquaculture-free harbor; the eastern sea area as a tourism zone; Tong'an bay as a limited aquaculture zone and Dadeng District as a deep water aquaculture zone. If the bay of Xiamen was a mess, how could Xiamen possibly become a port city? The Party congress in 2000 approved the proposal of building Xiamen into a modern port tourism city which was initiated by the State Council. Today, we have Haicang Harbor, cruise terminal, the Phase II, phase III and phase IV of Xiamen port passage. We have Xinglin Bay, which will become a picturesque residential district in the future. Now we have "one city with four supplements"—Xiamen Island as the main city, with Haicang new district, Jimmei new district, Tong'an new district and Xiang'an new district as the supplements. The eastern sea area is managed well and the construction of Xiang'an airport is well underway.

In conclusion, although we spent a lot of time and money, and although some leaders

▲ For the sake of environmental protection, environment experts believed that aquaculture should be eliminated; for the sake of economic development, fishery experts believed that aquafarming should be protected. After coordination by Xiamen Municipal Bureau of Oceans and Fisheries, two breeding areas were finally demarcated in Tong'an Bay. The picture shows the salt pans of Dadeng. (Photo / Lin Shize)

faced difficulties through the past 20 years of coastal management in Xiamen, we have brought many benefits to Xiamen and its citizens. Therefore, all of our efforts have been worthwhile. Today's Xiamen could not have developed without proper coastal management.

In terms of losses, they are likely due to the narrow-mindedness of some of our initial efforts. We didn't expect that Xiamen would develop so rapidly in these 20 years. For example, the policy for Tong'an Bay to develop aquaculture would have been better if it were "developing aquaculture with limitations within two five-year plans". However, generally speaking, this plan put forward by leaders and experts is suitable for Xiamen, as it is scientific and foresighted. The success of coastal management in Xiamen can be attributed to the hard work of young cadres, support from citizens and promotion by the media.

Based on my estimation, the cost of coastal management was over 200 million RMB. This amount of money may not sound like much today, but it was quite a large sum at that time. When I began my job as Vice Mayor, the general financial revenue of the whole city was only 2.8 billion RMB. In 2000, the revenue rose to 7.5 billion, and in 2001 when I started to work in provincial government, the number soared to 11.05 billion.

The busiest period of my career was the years when I took over coastal management in Xiamen. During those years, Xiamen was undergoing large-scale construction and development. I was the head of many working groups, such as the Health Insurance Working Group, Social Security Working Group and the Haicang Bridge Leading Group. No matter how busy I was, I would never slacken my work pace. At that time, I was the Executive Vice Mayor. As there were many holidays during the year, and workdays were limited, I wanted to seize each day and spare no efforts in working.

Through years of experience, the first and the most important lesson I have learned is to revere nature. This idea can be exemplified by the Wuyuan Bay case.

Wuyuan Bay was originally a harbor, and then became a salt factory. After reform and opening up, this area was used to breed shrimp, and the real value of the bay was overlooked. Later, opinions about managing Wuyuan Bay were divided into two groups: one group proposed to fill the bay in with earth; the other group to build a bridge. We thought that this bay could be developed into a yacht base. If it were filled, although this plan could have generated a large amount of revenue, the ecological environment would be destroyed. Unlike the area of the Conference and Exhibition Center, which was eroded by sea waves and could easily be filled back in, Wuyuan was a natural bay. To fill it in would be an offense to the gods! In addition, I believed it could become a yacht wharf after the bridge was constructed. In the future, it could be linked with Zhongzhai Bay by an underground tunnel, and then the water of Yundang Lake could be washed out, a difficult and expensive prospect if Wuyuan Bay were to be filled in. For these reasons, the government eventually adopted the second proposal: building a bridge.

The significance of policymaking can be demonstrated everywhere in coastal management. The decision of whether to open the Gaoji Causeway is an excellent

1
—
2

1. The sea and beach have become an integral part of this beautiful city. (Photo / Wang Huoyan)
2. Sailing competitions in Wuyuan Bay. All kinds of sailing events are held here. For examle, the China Club Sailing Challenge Tournament has been held for 11 consecutive sessions. (Photo / Wang Huoyan)

example. We ultimately decided not to open it for the following reasons: first, some people had strong emotional attachments to the Causeway. Second and more importantly, the government lacked sufficient funds. If the Causeway were to be opened, more issues would need to be coped with, such as flooding, traffic congestion and cross-sea electric cables. Generally speaking, the conditions were immature, so we decided not to open it for the time being.

Coastal management is a type of engineering project that requires persistence, wisdom and bold ideas. For example, for the construction of the Haicang Bridge, we needed an investment of 2.6 billion RMB. It was beyond the capability of the municipal government, so we needed to borrow money. At the time, the National Development and Reform Commission had a loan from the Japanese government which could be lent to Xiamen, but at last moment the deal fell through. I was unwilling to accept this fact and frequently requested that the leaders to give us the investment. Eventually, we were given new Japanese funding, which was something between a government and business loan with a long term and reasonable interest rate. These loans were settled, but we still lacked proper funding for start-up capital. I went to the China Development Bank to apply for the funds with real estate as a loan guarantee. They asked me which piece of land to use as a guarantee. In reply, I asked them, how much can you give me? They said 300 million RMB, which was a lot of money in the 1990s. I immediately accepted and said that I would use the Huoshao Island as the guarantee. My reasoning was that even in the event that we could not repay the loan, it would be acceptable for us to let them develop Huoshao Island, as Huoshao Island would always be in Xiamen.

I have lived with the sea for so long, yet two stories from before my career as water resources administrator stand out in particular. First, when I studied at Xiamen University in 1964, I went to Tong'an by ship from No.1 Wharf. At the time, when ships passed the Xiamen Causeway, there was a big hole that the ships could go through. Half a century later, the Causeway has been opened.

Another story is about my impressions of the blue sea of Xiamen: in 1989, I was told that the business tycoon Wang Yongqing was about to come to Xiamen to invest the Haicang area. One day at dusk, I went to Haicang by ferry together with the Secretary Wang Jianshuang and the mayor Zou Erjun. I saw dolphins leaping in and out of the water in groups. I said that this was a good sign, indicating that Wang Yongqing would come soon. Wang Yongqing never came, but we can still see dolphins of the Xiamen Sea leaping in and out of the water.

"Revering the Sea" in the Context of Urban Construction

Pan Shijian

Pan Shijian, former Vice Mayor of Xiamen, whose native place is Guangdong, was born and grew up in Xiamen, because of which his personality is similar with that of the sea: enthusiastic, open, tolerant, ambitious and profound in thought.

He once said, "The city is built on the sea, and the sea is integrated into the city", which is a fitting description for Xiamen. Yachts and sailboats, blue seas and clear skies, dolphins and egrets, beaches and waves are the achievements of 20 years of Integrated Coastal Management in Xiamen. He loves the sea and enjoys immersing himself in its beauty, feeling the sea breeze and inhaling the fresh salt air, experiencing a unity with nature described as "The sail, a single shadow, becomes one with the blue sky". Pan got tanned because of his frequent sailing and he joked that he was once mistaken for a farmer coming to work in the city. In the face of powerful tides and surging ocean waves, he can hardly find words to describe his emotions. However, the sea knows his reverence lies deep in his heart. Man cannot conquer nature; they can only conform to and show respect for it.

If we look through Pan Shijian's resume, we can find that he has lived and worked in Xiamen for his whole life, except for the time he spent as a student in Wuhan. He experienced the great changes brought about by the 20 years of Integrated Coastal Management in Xiamen. During this period, he, as the manager of city construction, was in charge of managing sea projects for 12 years. In his office, there was a map of Xiamen before the 1950s, on which Xiamen is an irregular U-shaped place. That beautiful and twisting coast line, which always puts him in a state of deep contemplation, has now turned into an oval. He knows that, as a result of 30 to 40 years of urban construction, large areas of the sea have been filled in, and on the surface, they seem tame, yet hide its roaring waves underneath. Therefore, he often introspects and reminds himself that protection of the marine environment is closely connected with the future and prospects of Xiamen as it is a coastal city known for the sea.

With even more rapid decline of the marine environment, coastal erosion and beach erosion are sending us a warning signal, which we should accept and understand in depth. Pan Shijian said it is a cruel truth that if you do not treat the sea with respect, the sea in turn will not accept you. He confessed that they have already realized, after the reform and opening up, that the sea is the lifeline of Xiamen and also a boon for its economic development. Therefore, the sea should be treated with care and exploited only cautiously. Once the marine environment is damaged, its restoration is far more difficult than that of land.

Thus, adhering to the principle of "endowing the ecological resources of the city to the public", Xiamen locals are making concerted efforts to develop the marine economy and construct a "marine society". Wuyuan Bay area, for example, used to be mud flats and fishponds, yet has been turned into a beautiful and modern new coastal area after over 10 years of construction. It is now the "new living room" of Xiamen, providing a successful example for the development and utilization of bay areas. Pan Shijian, an important proponent for the renovation of the Wuyuan Bay area, firmly believed that only through scientific and law-based governance of the sea, and through the strict pursuit of perfection despite all difficulties, can we endow the sea and the city with a true vibrancy and soul.

His body is with the jade ocean, his spirit the blue skies. Aspiring for high goals, he is accompanied by ships and inspired by sails. For Pan Shijian, it is not only a wonderful image, but also a lingering memory in his mind of a time when he was close to, in charge of and affectionate for the sea.

※ Increased tidal influx of 71 million cubic meters

Is urban construction a simple task or a difficult task? Should we only focus on the present, or should we take future development into consideration? The guiding philosophy for the development of Xiamen is that things done in the present should provide a better platform and environment for future development.

Let's begin with the construction of Wuyuan Bay. Located in the northeast of Xiamen Island, it is called "the place closest to Taiwan in Xiamen" by Taiwanese. Nowadays, droves of visitors are often amazed at the beauty and vigor of Wuyuan Bay. It has water areas of two square kilometers with rolling waves and glistening water; wetland ecology parkland of 89 hectares with abundant water, luxuriant grasses and flocks of birds (it is one of the best bird-watching sites in Xiamen Island). Additionally, it has an eight-kilometer path for electric scooters, an offshore leisure platform constructed of wooden trestle of 1.5 square meters; a pebble and sand beach of one kilometer; lush greenery and clean roads. Wuyuan Bay is not a place to hurry through, but rather a place to enjoy for half a day or longer.

It is hard to imagine that over 10 years ago, this place was a dull and strange area for most

1. From July, 1953 to 1977, seven Causeways including Gaoji, Jixing, Maluan, Yundang, Zhongzhai, Dongkeng and Dadeng were built. Causeway construction resulted in worsening seabed siltation.

2. Bridges that witnessed the development of Xiamen's marine rehabilitation. Thanks to these beautiful bridges, Xiamen, an island city, is also known as the "City of Bridges over Sea". The picture shows the Xiamen bridge serves as overpass. (Figure / Wang Huoyan)

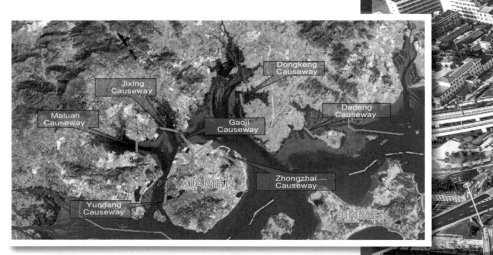

Introduction to the speaker

Pan Shijian, born in Xiamen, has served successively as the Deputy Mayor of Xiamen and the Vice Chairman of Xiamen Municipal Committee of the Chinese People's Political Consultative Conference (CPPCC). Since the beginning of his term of office in 2000 as the Vice Mayor of Xiamen, he has been fully involved in Xiamen's Integrated Coastal Management, engaging in such major projects as the withdrawal of aquaculture in eastern sea area, the integrated management of the western sea area, the integrated management of the eastern sea area, and the Causeways' Opening, the bay area restoration and the beach restoration. He is both the witness and the decision-maker of these projects.

▲ The rehabilitation plan for Wuyuan Bay: building bridges across the sea, cleaning up siltation, reclaiming land on both sides of the sea, opening up Causeways, and diverting water into the bay.

Xiamen locals. Wuyuan Bay used to be a beautiful sea bay. However, it was transformed into a salt factory when the people's commune built a dam in order to solve local people's difficulties in life. After the salt factory stopped production, the bay deteriorated into mud flats. Looking at the picture of Wuyuan Bay over 10 years ago before renovation, we find it was occupied by mud flats with no sign of a sea bay. This place was considered as unlivable for its being sparsely populated and located in a draught area with serious coastal erosion as well as large areas of mud flats inside the bay, which stretched to or even over the middle line of the sea. However, historically, it used to be a beautiful sea bay.

How was the place to be renovated?

It is possible to build another Causeway and flood the mud flats. The initial goal of this program was just to widen the Zhongzhai Causeway, as it was included in the Huandao Road which had to be paved to the airport. At the beginning of renovations, this sea bay was enclosed for aquaculture, of which the local government and experts held contrasting

▲ After rehabilitaion, Wuyuan Bay was recognized as "the New Living Room of Xiamen", "the New Pearl in Haixi", and "a paradigm of beautiful Xiamen in the 21st century". (photo / Wang Huoyan)

opinions. The first opinion was filling in the salt pan and marshlands to construct buildings because of the high value of the land. However, if we chose commercial construction, we would lose this beautiful sea bay forever, decreasing sea area and tidal influx. The second option was to open a Causeway and convert cultivated lands into an ocean area so as to build "a beautiful living room" for this city on the sea again and restore its erstwhile beauty. Moreover, it would increase tidal influx and greatly improve the marine ecological environment.

To weigh the advantages and disadvantages of the proposals, we held many workshops with experts and finally came up with the principle of "ecological protection and endowing the local resources to the public". We opened the Causeway, dredged the middle area, reclaimed land on both sides, built bridges across the sea and led water to the bay, creating a space for close communication between people and water. In the renovation of the inner bay of Wuyuan, instead of using traditional methods to isolate the inner bay from the sea by building dams on the periphery, we interlinked Wuyuan bay and the East China Sea. In the leisure renovation area around the bay, instead of just filling the inner bay, we followed the three steps of intercepting sewage, dredging and returning fishponds to sea. Through dredging and hydraulic reclamation, we built a land area of three square meters on both sides of the bay as well as exploded the reef and dredged the outer bay to minus seven meters' water depth. Our experience showed that these two new land areas did not decrease the tidal influx; on the contrary, they increased tidal influx by 250 million cubic meters and constructed coastline of 7.5 kilometers, greatly lengthening the bank line of Xiamen.

During urban construction, we always adhered to the principle of leaving the best land to our citizens and our culture regardless of their social status and money, which was also the principle guiding the renovation of Wuyuan Bay. Over the wetlands park, there is the Hubian reservoir, whose water pass through the park and run to the sea. We planned to reserve wetlands and return the bay to the city and citizens so that when they walk out of their homes to this place, they may enjoy mountains, water and thousands of precious birds inhabiting here. After the Causeway was opened, the sea bay reappeared and in it a rare wetland was reserved. There has been a beautiful urban transformation in the Huandao Road from Xiangshan Mountain to Wuyuan Bay (previously Zhongzhai Bay).

To be honest, I visited some places where local people were rolling up their sleeves preparing to fill the sea. I wondered why they filled such a beautiful sea and advised them to open up the sea bay, yet they said that Xiamen opened the sea bay only because it was rich. Actually, we did this not for the reason that Xiamen was rich, but because if we built the Causeway to increase land area, we would lose many industries and destroy the biological chain. Therefore, though it took a large amount of money and did not gain support from many people, we still decided to open the bay after consulting advice of all parties. We also learned from the renovation of Yundang Lake. It was an open sea bay in history and later was turned into inner lake, whose wonderful scenery can no longer be restored, being perhaps the pity of Xiamen. Xiamen spent 30 years and a large amount of money to fill Yundang Lake, however, it has not been completely rehabilitated even today. Therefore, treating the environment with respect cannot be stressed hard enough.

Xiamen is an inner-bay coastal city. Thus, only sufficient tidal influx can ensure the health of the marine ecosystem. The sea is a good natural cleaning system and water flow can produce and retain life. Just as rivers must keep flowing, otherwise they will stagnate, sea water needs constant exchange, providing fresh sea water at high tide and taking dirty water away at low tide. If this place were filled, the water level would drop and dirty water would increase. After the Causeway was opened, the entire Wuyuan Bay increased its tidal influx by more than 5 million cubic meters and the whole city's tides increased by 71 million cubic meters, an invaluable service to the city, equivalent to that gained by digging out half of Xiamen Island.

※"Heaven, earth, sun, moon and people" is the fundamental nature of Wuyuan

After nearly 10 years' development and construction, Wuyuan Bay becomes more beautiful and a more desirable urban environment with a newly-built 2-square-kilometer inner bay, 3-square-kilometer land and 8-kilometer-long high-quality coastline. When the Xiamen International Airport is relocated to the Dadeng island in the future, this beautiful bay, which was once surrounded by Causeway and later was reopened and dredged, will be located between the future airport and Xiamen city center and become the city's new "living room".

This "living room" is for everyone, not only for a few. Therefore, in urban planning, we did not simply divide the land and sell it to land developers. Instead, the government beautified public buildings and public spaces in the early stage of development. The creation of a beautiful ecological environment is only the first step. The highlight of Wuyuan Bay is the integration of construction and culture.

In order to fully embody the concept of ecological culture in bay scenery design, five circular arched bridges are lined up and these arches, with their shadows, form circles at full tide, becoming a unique iconic feature of Wuyuan Bay. When our ancestors built the city of Beijing, they built the Temples of Heaven, Earth, Sun, Moon and the Forbidden City in the middle. Only under the protection of heaven, earth, sun and moon did the emperor dare to enter the Forbidden City. Xiamen inherited this concept from our ancestors in city construction, therefore, these five bridges were named as "Tianyuan Bridge", "Diyuan Bridge", "Riyuan Bridge", "Yueyuan Bridge", "Renyuan Bridge" respectively after the heaven, earth, sun, moon and people. These names imply that people live between the heaven and the earth and that man and nature are one. They also warn people that man cannot conquer nature, they can only conform to and show respect for it. Yuan, which means "full" in Chinese characters, was renamed Yuan, which means "relationship". The meaning of the former character is included in the later one and the later character also implies that a harmonious relationship between nature and people can be called a full one.

Wuyuan Bay area has a wetland of 89 hectares, which is vividly described as Xiamen's unique "green lung". The wetland has freshwater as the major water sources and abundant typical island vegetation, and is a suitable habitat for birds, thus home to about 54 species of birds. Therefore, during the renovation of Wuyuan Bay, it was important to conserve the wetland park and protect the natural features of the wetland. In the development of wetland parks, builders

tried their best to maintain the ecology of freshwater wetlands, highlighting their function as one of the best bird watching spots in Xiamen, rehabilitating and developing existing spillways, freshwater lakes, pastures and farms, setting unmanned core protection areas on the south bank of freshwater lake and creating a paradise for birds. At the same time, they also built the central park of the city where people live in harmony with nature through such features as wooden trestles, pavilions, terraces, towers, causeways and arch bridges. Walking in the wetland park, people may enjoy exuberant grasses, flying birds, pavilions, terraces, towers, distant mountains as well as integrated light and colors of the lake. There is a wetland maze in the bay, where children and the elderly can enjoy themselves. Small kids will find the maze most amusing—though the exit seems just around the corner, the

▲ Sailing, a sculpture, and sail boats in Wuyuan Bay

participants simply cannot get to it.

Xiamen also introduced a lot of new birds to Wuyuan Bay, leaving the best land for nature, animals and life. Take the cormorant (Phalacrocorax carbo) for example. It is a migratory bird. Every fall, thousands of cormorants fly from Qinghai Lake and over our heads in a line. In the past, they had their bedroom in Jinmen and their eatery in Xiamen, flying from Jinmen in the morning, feeding around the waters of Xiamen, Yuanboyuan Expo Garden, Yundang Lake and Dadeng Island during the day and flying back to Jinmen in the evening to spend the night. Now that the environment in Xiamen, including Wuyuan Bay has been improved, many cormorants gather in Xiamen. Therefore, Wuyuan Bay Wetland Park not only provides a place for people to play, but also embodies the harmony between nature and people. If there were only people, cars and buildings on earth, it would be too terrible, almost a hell. Only when endowed with nature can it become a paradise. So many birds live with us and we do not disturb them, you see, this is paradise, isn't it?

Wuyuan Bay is also a paradise for the blue-cheeked bee eater (Merqps superciliosus), which is a ferocious bird under beautiful and elegant appearance. Reputed as "China's Most Beautiful Bird", these birds, whose ancestral home is Xishuangbanna in Yunnan province, fly all the way to Xiamen in order to brood, and their only breeding place in Xiamen is Qima Mountain in the bay. This mountain is special because birds can drill holes into the rocks, enter into these holes and lay their eggs. The dense caves are like honeycombs. Many people did not know that there were such birds in the mountain, so excavators were driven to the nesting areas and the hills were nearly leveled. We received letters from citizens and had a meeting with the contractor overnight. We insisted that the mountain should be protected from excavation as once the mountain was leveled, we would never again see these beautiful birds. We can build two fewer buildings, but cannot deprive China's most beautiful birds of their home. Finally, legislation was resorted to so that the bee eater can live here freely.

▲ Chestnut Bee

▲ Wuyuan Bay Wetland Park

With the completion of Wuyuan Bay's ecological renovation, it becomes the starting point of the Xiamen Marathon, a sailing center and also home to the Wuyuan school village which combines music education with modern education, a shining example in Xiamen of modern education. This village consists of primary, secondary and special schools (Xiamen No. 2 Middle School in Gulangyu moved here; Xiamen Municipal Government and the Central Conservatory of Music co-founded Gulangyu Piano School in Wuyuan Bay). Its main principle is to bring together students of music with students of general nine year compulsory education in a large school village, so that ordinary students can be close to children studying the arts. For one thing, art school teachers can give classes to general school children, exposing them to art. For another, those music school students, who spend most of their childhood in the practice room, can play happily with the other children after class.

The Gulangyu Piano Museum exhibits more than 200 pianos of antique and modern construction, some Chinese and others of foreign origin, collected by Huang Sanyuan, a

▲ An open air plaza of music by the sea

piano tuning master. Each piano carries with it with the unique sound of its time. At present, the museum has become an important venue for primary and secondary school students to have their extra-curricular activities. In order to enrich local culture, we have also given cultural exhibitions, built an outdoor music square and established a celebrity sculpture plaza, in which 30 statues of celebrities, scientists, musicians and artists from all over the world are placed for us to show them our gratitude for their contributions to mankind. When future generations enter this beautiful bay, they may feel like they are walking into history and exploring the greater world.

Conserving the best land for future generations is a way of showing respect for the public and also a means of cherishing culture. Zhongzhai Village, where Wuyuan Bay is located, is a village inhabited by She ethnic minority group. We promised to build a sculpture for the ancestors of Zhongzhai Village during construction. At first, the elderly in this village did not quite believe it. However, builders now have set up a large-scale sculpture named "Sail", conveying a strong nostalgia of local fishermen and symbolizing that ships in Wuyuan Bay will set sail for the bright future.

※ Turning the rot mud flat into "the West Lake of Xiamen"

Now, let me introduce the Yuanboyuan Expo Garden. It used to be an important bay, but later, to allow for the construction of the Yingtan-Amoy Railway, it was enclosed by the Jimei Causeway. At that time, there was little marine pollution in Xiamen thanks to its relatively small population. Thus, given the conditions then, the construction of a single Causeway shouldn't have caused any problems. The innovative gentleman, Tan Kah Kee, even made a proposal to build a tidal-power generator alongside the Xinglin Causeway.

The bidding for the China International Garden and Flora Exposition project in 2007, including the process of site selection, planning, construction and operations, was in accordance with the theme of "Harmonious Coexistence, Inheritance and Development". However, the main purpose for building the garden was to provide the citizens with an ideal location for relaxing, sightseeing, and exercise. The site selection process was not easy. There were three choices: expropriating land in Xiang'an, demolishing existing residential buildings in Tong'an, and building on Zhongzhou Island of Xinglin Bay.

After thorough consideration, we selected Zhongzhou Island, leaving the farmland and urban spaces as they were. "The grand expo on water" was to be built on a large mud flat through reclamation by pump-filling. The planned area for the Yuanboyuan Expo Garden is 6.76 square kilometers, including 3.03 square kilometers of land area and 3.73 square kilometers of water area. At that time, many people thought it would be too difficult to build on Zhongzhou Island, especially within the one year timeframe.

Why was it so difficult?

At that time, there was a grid of fishponds and shrimp ponds in Xinglin Bay. After dozens of years of land reclamation for aquaculture by the residents, two major problems emerged. First, long-term aquaculture had caused deterioration of water quality and serious environmental damage. Second, owing to the lack of proper management, the storage capacity of Xinglin Bay continued to shrink and its capacity for flood storage and flood control declined. When storms hit, flooding often followed, threatening the safety and properties of nearby residents.

The marine environment would benefit if we built the Yuanboyuan Expo Garden in Xinglin Bay, but we should take care not to cause secondary pollution in construction. During construction, Xiamen emphasized flood prevention, draining stagnant water

1/2. The Garden Park was full of grids. These grids were, in fact, large and small fish and shrimp ponds.

3. The Garden Park did not occupy cultivated land or urban development land. Instead, it was built on a large mudflat, through blow–flushing, in the form of moon–shaped garden with flowing water. (Photo / Wang Huoyan)

and environmental protection. After renovation, the water area has reached 6.9 square kilometers, which greatly increased the capacity for flood storage and flood control of Xinglin Bay. At the same time, Xiamen also focused on ecological impacts and effects on the landscape. Before construction of the Yuanboyuan Expo Garden, we surveyed the local animal and plant species such as birds and fish, and took samples of the water quality and aquatic biota to make sure that we would not break the ecological balance and would maintain the original ecology. We have tried our best to maintain the original green spaces, water system, and vegetation to achieve ecological benefits with smallest intervention. To best maintain the original landscape, we worked to balance the earthwork on the site to avoid large scale terraforming. During the construction, reclamation was carried out only on the original revetments to protect the revetment aquatic plants.

The newly built Yuanboyuan Expo Garden consists of 11 ecological islands, approximately

30 kilometers of coastline, more than 1,500 species of plants and 98 species of birds. There are abundant terrestrial and aquatic plants and abundant fish resources. Built on the water, the Garden contains nine scenic isles and one hundred smaller gardens, presenting a picturesque combination of exquisite craftsmanship and natural beauty. The 12-square-kilometer water surface of Xinglin Bay and the newly-formed large area of green land and wetland serve as a green protective screen to relieve the urban heat island effect, regulate the regional climate and optimize the living environment.

In terms of the green protective screen, Xiamen has taken lessons from the restoration of coastal mangrove wetlands. First of all, "mangroves" refer to the woody plant communities that grow in the intertidal zone of the estuary. They fight at the frontier to block the tide and the sludge, protecting the living space of the creatures in the mud flat. In the past, we didn't know that mangroves were such an integral part of the marine ecosystem. It was not until we realized that the mangroves could protect the land and the sea, purify sea water, and improve biodiversity that more efforts were devoted to the management of mangrove forests. At present, the mangrove forest at Xiatanwei, Xiang'an district is the first mangrove-themed wetland ecological park in Fujian Province, where the mangroves grow quickly and have reached the height of a man.

On the subject of beach restoration, Xiamen also boasts an experience worthy of emulation. In beach restoration, sand selection is very important. Fine sands will be eroded quickly by waves, while coarse sands prevent people from walking on the beach. Scientific corroboration is required to restore the beach to its natural state.

During the process of urban construction, Xiamen has saved all of its beach, sea and sunlight resources for the people. No coastline is exclusive to privileged individuals. During construction, the cultural and historical links have been taken into consideration so that the seaside and the lakeside areas are preserved rather than sold to real estate developers.

There is a Chinese Education Park in the Yuanboyuane Expo Garden, the first Chinese Education Park in China. To build it, I even went to Beijing to ask the Ministry of Education for approval. It is fair to say that the list of the best Chinese scholars in history that we have selected for the Park has won the approval of the Ministry of Education.

The entire Chinese Education Park covers an area of around 20 hectares. The park presents China's scholarly history with vivid displays and close-ups of the key figures and history. It consists of six minor gardens with the themes including classic scholarly scenes, stories of the great sages, an overview of the history of education, the establishment of elite schools, educational methods and international educational exchange. The large number of ancient books stored in Xinglin Academy were all donated by citizens.

Every bridge displays its unique spirit through its name. Each of the 15 bridges in the Yuanboyuan Expo Garden has a name informed by Chinese culture, including Tansi Bridge("musing"), Qianxiang Bridge ("artistic empathy"), Yunzhen Bridge("treasure"), Ninglv Bridge("meditation"), Danbo Bridge("composure"), Yuanrong Bridge("fusion"), Yiyuan Bridge("broad-minded"), Zuyin Bridge("hospitality"), Qingyuan Bridge("lofty"), Haojie Bridge("clean"), Kongming Bridge("intact"), Yinwei Bridge("subtle"), Chengming Bridge("bright and clear"), Ruizhi Bridge("wise") and Minwu Bridge("enlightened"). They all have different shapes, through which we want to exhibit the classic bridge styles from home and abroad.

▲ The white dolphin plays in Xiamen sea area.

※ Never give up the concept of marine ecological protection

Residents in Xiamen have deep attachment to a group of so-called "white angels". Many people make a special trip to the waters outside Wuyuan Bay, around Jiyu Island, Houyu Island, Dayu Island, and Huoshaoyu Island of the western port and nearby Xinglin Bridge just to watch them. You may have guessed that these angels are white dolphins. Today, Xiamen is not only one of the most important habitats for white dolphins, but also the only city in the world with white dolphins visible in urban areas.

Let's start with the history.

Before the 1980s, Chinese white dolphins were spotted everywhere in the sea around

▲ Children and the elderly shouted "the train is coming", unfolding a new chapter for travel through the first underwater tunnel, Xiang'an Tunnel, in the country. (Photo / Li Wanhan)

Xiamen. It was quite common to see the spectacular scene of the playful white dolphins chasing each other and leaping out of the water. It may not occur to many people that where there are dolphins, there are no sharks, because dolphins are cooperative and will fight against sharks together. After the 1980s, due to construction in coastal areas, the disorderly exploitation of sea sand, excessive aquaculture, the deterioration of water quality and illegal electric explosive fishing, the dolphin lost its ideal habitat and as a result its number dropped sharply.

Fortunately, after improvements to the marine environment through dredging, the white dolphins have come back, flocking in groups. When I drive a sailing boat alone sometimes, I will come across a couple of white dolphins swimming near to greet me. Sometime around 2008, a person caught a precious scene of six white dolphins in a single picture in Wuyuan Bay. In such an inner bay as Xiamen, it is almost a miracle to see six white dolphins at once. Some people do not know what the dolphins look like. One person looked at the photo and found it very strange, asking if a boat had turned over and several pigs had fallen into the sea.

Improvements to the white dolphins' habitat are in large part thanks to the promotion of marine ecological restoration by the Xiamen municipal government. The opening of the Causeways was the prelude to large-scale restoration. The historical contributions of Causeways are indelible, and have turned a deep chasm into a thoroughfare, and changed an island into a peninsula. But for the Causeways, the Yingtan-Amoy Railway could not have reached Xiamen, and could not have helped it win its position through the benefits brought by convenient transportation. Beginning in the 1950s, in order to solve the basic living problems of the people, Xiamen enclosed tideland for cultivation and used reclaimed marine land for the development of fisheries and the salt industry. From 1953 to 1977, the city built seven Causeways including Gaoji, Jixing, Maluan, Yundang, Zhongzhai Dongkeng, and Dadeng, to promote economic development and expand the construction of Xiamen Island. However, the construction also gradually reduced the capacity for water exchange of the Xiamen sea area. As a result, it came as a shock when silt appeared in the sea that used to be clean.

The Causeways separated the eastern and western water areas and blocked the channels of the ships and the dolphins. The sea area of Xiamen gradually became a semi-enclosed bay, weakening its water exchange capacity. Due to the deteriorating water quality caused by aquaculture, the pressure on the environment increased and silt on fairways and shoals accumulated, affecting the capacity for navigation and sustainable development of the port. Given the historical and scientific conditions at that time, it was impossible to protect the marine ecology while building the Causeways. However, today we are able to maintain sustainable development alongside construction.

The transformation of the seven Causeways was carried out at the same time. Four of the seven Causeways were transformed into bridges. But the other three couldn't be demolished, because if they were removed, the nearby road surface would be flooded at high tide. So, what was to be done? In the end, we chose to increase their storage capacity of tidal water by building water gates for flood prevention. Among all the projects, the final plans for the transformation of Xinglin Bridge and Xiang'an Tunnel both made space for the protection of white dolphins.

Xinglin Bridge passes directly through the core zone of the conservation area for white dolphins. Located within the conservation area according to the *Environment Impact*

▲ *Looking in the mirror,* a sculpture by the sea.

Assessment Report, the construction of the bridge went forward with no explosive reef demolition and no reclamation of sea land. The dispute over the project of Xiang'an lay on whether to build a bridge or a tunnel. Building a tunnel would cost six to seven million more RMB than building a bridge. At the same time, the construction of a tunnel would involve more technical problems and risks. It was a tough decision, but Xiamen made the choice for the sake of marine ecological protection and chose to build a tunnel. One of the most important reasons was that the planned construction was in the conservation area of China white dolphins, which meant that the explosion during bridge building would threaten the survival of the white dolphins. White dolphins are very sensitive and usually don't swim through rocky areas, so the piers under the sea like, big machines, would drive away the dolphins.

We spent a long time running calculations and simulations by mathematical models and physical models and finally overcome all the technical problems. In terms of water pressure, one square meter of the tunnel surface is under the pressure of 70 tons' water. If a tiny mistake let in the water, it wouldn't be a small problem of leakage. Instead, the water would shoot into the tunnel like bullets and slaughter all the passengers. It was in the spirit of caution rather than adventure that we succeeded in building the first sub-sea tunnel in China within five years without a single fatal accident. This project was listed as one of the top ten most dangerous projects in China.

On April 26, 2010, Xiang'an Tunnel was formally opened to the public. The opening was simple, yet unique. Under the sculpture named "Never Admit Defeat" at the Wutong End of Xiang'an Tunnel, six citizens including five children and one senior cheerfully announced "Open to traffic!" The senior was an ordinary local man, representing the history of the area and expectations for the future, and the children represent the hope. The senior raised one hand and said "Open to traffic!" in the local dialect, welcoming the special moment in a unique way. There are reliefs at the entrance and inside the tunnel displaying the scenes of the years of hard construction. There is a relief of 70 meters in the tunnel named "Never Admit Defeat" in remembrance of the ordinary workers who never gave up despite the tough work conditions and made great contributions to this city. It also reminds us to never give up in protection of the environment and local ecology.

※ Creating a blue economic industrial chain of hundreds of millions

The extraordinary experience of Xiamen is thanks to the balance of economic development with marine ecological restoration to achieve harmony between man and nature.

Among the places most in need of renovation, the Jiulong River was so severely polluted that the waters of Xiamen were all covered with water lettuce in heavy rain. If not cleaned, it would create serious problems for Xiamen. Jiulong River is in southern Fujian Province with a total length of 258 km, flowing through Longyan, Quanzhou, Zhangzhou, and Xiamen. Its basin area reaches 14,000 square kilometers, and finally flows through Xiamen Bay into the East China Sea. Along the banks were major sources of pollution including paper mills, sugar refineries, pig farms and urban sewage. Most of the water pollutants entered the Xiamen sea, accounting for about 80% of the total amount of land-sourced pollutants.

At present, Xiamen, Zhangzhou and Longyan have launched joint investigations and established the Xiamen Port Authority to promote the comprehensive improvement of the Jiulong River. To be honest, the founding of Xiamen Port Authority is more favorable to Zhangzhou and Longyan than Xiamen itself, who just provides services. However, Xiamen still chooses to shoulder the responsibility for three reasons: first, Xiamen people have a deep concern of the future protection of the ocean; second, Xiamen is aware of its duty and is willing to communicate and cooperate with its neighbors despite compromising part of her profits; last but not least, this authority will have a positive influence in promoting law-based protections.

Xiamen's sea areas will play an important role in its future development. Therefore, it requires proactive action to promote its protection and at the same time encourage innovation of new industries. Xiamen has decided to promote the marine economy as an industrial chain of hundreds of millions of links through the strategies of "bringing in" and "going out".

In the past, people did not attach much importance to marine industry. There were only two models for marine industry, fish or shrimp farming and shipping. When we went abroad and saw the beautiful oceans of other countries that breed countless industries closely related to daily life, for the first time it occurred to Xiamen that we could develop the marine economy in our own blue territory. Facilitated by the natural advantages of this city, Xiamen has promoted cruise, yacht and sailboat enterprises, and Wuyuan Bay now serves as a center for yachts and sailboats.

In accordance with national directives, Wuyuan Bay is aimed to be one of the best yacht harbors in China, one of the harbors with most multi-million yuan yachts, and one of the best organized sales bases for yachts and sailboats on the southeast coast. Now sailing in Xiamen is

▲ The Map Square in Wutong Lighthouse Park, Xiamen, cleverly integrates marine culture in the park. (Photo / Taohui)

at the forefront of the country. It has been my pleasure to witness the growth of the sailboat industry and its popularity among ordinary people. The long queuing beside the harbor shows that this activity is no longer exclusive to a few rich people but has started to benefit the public.

When we protect the ocean, the ocean will reward us and its rewards are far-reaching.

Every year, there are activities of various kinds to promote maritime culture and to promote the scientific understanding of the ocean on the "World Ocean Day", "China Ocean Day" and "the National Marine Disaster Prevention and Mitigation Day" as well as in "the World Ocean Week". The fruits of the "World Ocean Week" every year have a profound impact on the world, showing the determination of the Chinese people to protect the oceans, develop the marine economy, create a marine culture and strive to be a maritime power. At the same time, Xiamen hosts annually the International Yacht and Sailboat Race, the College Students' Sailing Race, the Cross-Strait Sailing Race and other competitions and activities. Importantly, the sailboat named after "Xiamen" successfully has completed its voyage round the world, the first to finish the journey in China's sailboat history. This cruise has carried forward the motivated marine spirit of Xiamen, and has raised the people's awareness of the sea.

The theme of the 2017 BRICS Gala is Sailing to the Future. The Gala started with the sea and connected its chapters with the sea, telling the world a story of the sea, which reminds me of the song "Dream of the Sea". It was composed by the famous musicians Yin Qing and Wang Xiaoling for Xiamen, but not just for Xiamen. It is also a great representation of the ideals of today's marine culture. The lyrics are as follows:

Since the ancient time of sailing
Our dream of you has never changed
Flying on the wings of the white egrets
Hovering over thousands of waves flowering on the blue ocean
Dashing into the ocean waves with our hearts open
Desiring to be prosperous and strong
Dreaming for a better world with peace and love
We are striving towards a brilliant future
See the silk road circling and extending like colorful ribbons
Our common fates are linked by our blood
Let us go sailing along with the song of spring winds
And create new legends throughout the world
Dashing into the ocean waves with our hearts open
Desiring to be prosperous and strong
Dreaming for a better world with peace and love
We are striving towards a brilliant future

The Xiamen Model Is of Epoch-Making Significance in the History of World Coastal Management

Chua Thia-Eng

▼

For the locals in Xiamen who live by the sea, the name "Chua Thia-Eng" carries significant weight, for it is him, a coastal managing expert from Malaysia who brought advanced ideas to Xiamen over 20 years ago and thus promoted the launching of Xiamen's integrated coastal management initiatives. Chua Thia-Eng and Partnerships in Environmental Management for the Seas of East Asia (PEMSEA) contributed to the changing land- and seascapes around Xiamen from an "disorderly, competing, and unsustainable coastal development dilemma" into an indispensable source of vitality, towards transforming Xiamen into a "Garden City by the sea" that has spurred Xiamen's rapid economic, social and environmental development into a sustainable city.

The 77-year-old Chua Thia-Eng was born in a small Malaysian village. He was educated at Nanyang University and the National University of Singapore and received postgraduate training at the University of Tokyo. After graduation, he worked at the National University of Singapore, the University of Science, Malaysia and the Food and Agriculture Organization of the United Nations (FAO). Thanks to his days spent in East Asian countries, the sea was no stranger to Chua Thia-Eng. He was involved in several international marine projects that brought him in greater contact with the issues related to coast and ocean management. These experiences made this him an expert and leader in comprehensive and integrated coastal and coastal management, thus earning him the well-recognized title the"Father of Integrated Coastal Management". People say tha"A prophet is not without honor save in his own country", but Chua Thia-Eng's work in Xiamen was never easy, for his ideas had never really been tested in other places before he came to Xiamen.

"Integrated Coastal Management is a local government-led, top-down, cross-sector, interdisciplinary management approach, the implementation and sustainability of which is an unprecedented daunting task. I myself and many international initiatives have made several attempts, but all encountered difficulties to sustain the management initiatives. Finally, we succeeded in Xiamen. I think in this process 'people' and governance factor play a decisive role. " According to Chua Thia-Eng, the experiment in Xiamen was just one more attempt after many failures of demonstrating the effectiveness and sustainability of coastal management. He never expected that Xiamen's Integrated Coastal Management initiatives could be so successful. "The success of Xiamen model not only benefits the future generations of Xiamen, but also brings valuable reference for other countries and regions as well. This is of epoch-making significance in the history of coastal management in the world. "

※ Xiamen was selected because of its special autonomous status

In 1992, I led a UNDP project developing team to explore and develop a Global Environmental Facility (GEF) Large Marine Ecosystem's Project on marine pollution prevention and management for the East Asian Seas region. Hence the opportunity to visit several countries bordering the seas of East Asia.

At that time, marine pollution was a "hot" environmental issue in the world, but a tour around East Asian seas region and consultations with governments and scholars convinced us that it was a wishful thinking to effectively resolve marine pollution problems once for all. At that time, during the designing of the regional programme on marine pollution prevention and management, we intended to include a special project to apply comprehensive and holistic management approaches to resolve marine pollution problems encountered in the region. Unfortunately, this will involve huge investments in terms of human resources and capitals. UNDP alone could not undertake

Introduction to the speaker

Chua Thia-Eng, Chair Emeritus, Partnership Council, the Partnerships in Environmental Management for the Seas of East Asia (PEMSEA), member of the International Ocean Institute Governing Board, Senior Adviser to the Institute of Ocean and Coastal Development of Xiamen University. In 1994 he introduced and promoted an advanced concept of coastal management in Xiamen—unfolding a new chapter of integrated management of Xiamen's coastal area. He was widely recognized as the "Father of Integrated Coastal Management. " In 2011. He was awarded the title of "Honorary Citizen of Xiamen".

▲ Sunset in Gulangyu (Photo / Lin Shize)

such an endeavor. A bold idea thus emerged—start from local initiatives by addressing marine pollution challenges within the management boundary of local governments and subsequent geographical scaling ups. In other words, managing coastal issues under the framework of local governance and managements in close partnerships with the stakeholders.

This regional programme unintentionally laid the foundation for the establishment of a long-term regional body, the Partnership in Environmental Management for the Seas of East Asia (PEMSEA)!

In 1993, the UNDP/GEF team began its dialogue with the State Oceanic Administration of the People's Republic of China. Back then, China was still in the early stages of "structural reform and open door policy". Many existing foreign-aided projects in the country were still at the pilot stage and mostly operated directly under the central government. But the fate of this project largely hinged on whether the selected local government could be authorized to manage the project by itself and take control of its available human and financial resources. After negotiations, the Regional Program, was able to secure the central government's approval and support for Xiamen Municipal Government to take full responsibility in the implementation of the integrated coastal management project.

There are many coastal cities in China, so why Xiamen? This decision was a result of many factors. At that time, the State Oceanic Administration recommended two cities for our consideration: Xiamen and Qingdao. As the southern and northern centers for marine research in China respectively, these two cities had strong scientific capacity which are perfectly important for developing science-based management programs. As Xiamen had just been designated as one of the five Special Economic Zones by the central government, it will have greater local autonomy and therefore ideal for the implementation of external aided projects such as this one. Our main concern was to ensure the development of a successful coastal management model that can be replicated in other coastal cities in the country and abroad.

Consequently, we selected two cities—Xiamen, China and Batangas Bay, in the Philippines as demonstration sites for implementing Integrated Coastal Management Programs. These two cities had completely different political systems. China was a socialist country, and success in Xiamen meant possible application to other socialist countries like the DPRK and Vietnam. The Philippines, on the other hand, was a capitalist/democratic country, the success of demonstration site would contribute to replications of the approach to other countries in the region where decentralization had already occurred, thus creating favorable conditions for local implementation of coastal management initiatives.

In the 1980s, I was involved involved in the FAO's regional aquaculture training program in Asia and the Pacific and thus had the opportunity to tour the coastline of China. I was most impressed by the seas around Xiamen. At that time, the gun fire between Xiamen and Kinmen had just ended several years ago. The city administration was eager to initiate economic development, undertake structural and management reforms whilst opening up to external economical, cultural and scientific co-operations. With its 1.2 million people, blue sea and warm sea winds, Xiamen reminded me of Penang, Malaysia.

So, personally, I hoped that this project could be launched in Xiamen and help Xiamen preserved its clear water and blue skies by introducing integrated coastal management concept and approaches. However, it is worth noting that however hard we tried, if the Xiamen government was not interested, we would not be able to proceed. Therefore, the final success of this project depended largely on the attitude and political commitment of the Xiamen Municipal Government.

After gaining authorization from the central government, we went to negotiate with the Xiamen Municipal Government almost immediately, and were glad that the local government showed great interest in this international project. As a matter of fact, we had not expected such an enthusiastic response, but this response had added to our confidence in the project's success. Therefore, in 1993, we reached an agreement to launch the project in Xiamen in early 1994.

※ Coastal management in Xiamen continued to gain importance and support from various succeeding leaderships

In 1994, Xiamen had already begun rapid economic development including infrastructure reconstruction. Many dredgers were operating incessantly in the coastal and sea areas, resulting in huge damage to the coastal land and seascapes including negative impacts on the ecosystems. This was a pressing issue facing us after we started the project. However, change could not come solely from foreign help, but had to rely on local government efforts. The top priority then was to increase the awareness, knowledge and capacity of local government officials so that the local government could solve the problem on its own.

▲ Listen! It's the sound of Xiamen's Sea. (Photo / Cai Mingde)

▲ On November 24th, 2009, PEMSEA (Partnerships in Environmental Management for the Seas of East Asia) awarded Xiamen Municipal Government the highest award in East Asian coastal zone management: "Outstanding Achievement Award for Local Government in Sustainable Development of Coastal Zones".

Why were we so sure this could succeed? In the past 20 to 30 years, not only PEMSEA but also many other organizations in the world had attempted to manage and control marine pollution in the region, but none had made significant progress. What often happened was that when the expert team left and the start-up capital ran out, the project would end. In the 1980s, I once implemented an USAID-funded Coastal Resource Management Project in six countries in South East Asia, but failed to reach plan implementation stage. The project was totally led by scientists. We wrote many books, reports and journal articles, as well as organized many regional scientific meetings. People were very excited at first, but after we completed the project and left the scene, there was no follow-up actions. The biggest mistake here was that we failed to involve local government and stakeholders in project design and implementation. So, this time, in Xiamen, we tried to introduce the concept and approach of integrated coastal management to the local governments and involved them in project design and implementation—this was the first time we had done so and totally in uncharted waters.

Integrated coastal management not only covers wide geographical areas but also involved many different economic sectors and management challenges with political, social and environmental consequences. First, we needed to gain the confidence of major officials in the Xiamen Municipal Government so that they could lead and promote the project implementation. Second, the Xiamen government had to strengthen its working relationship with stakeholders' organizations such as people organizations and NGOs, so as to build working partnerships in the development and implementation of integrated coastal management. Third, integrated management approach is process- and capacity-based and hence require a longer term program approach—it would need several year or decades to achieve sustainable development targets, despite the fact that the leadership in the municipal CPC committee and local government changed every few years. How could we make sure that the new leadership would carry on where their predecessors left off and continue to value integrated coastal management? These were the problems facing us in the initial stage.

Several adjustments had to be made.First, this project was initially classified as a scientific project and was assigned to operate under the leadership of the Xiamen Science and Technology Committee. However, the Committee had no jurisdiction over the entire project implementation, as many aspects of the project fall under the functions of other agencies, so a higher level agency would be needed to take charge. In the end, it was decided that the project would be best operated under the direct control of the municipal government while the Xiamen Science and Technology Committee would continue to lead the project. This decision posed some difficulties, because in China the usual practice was for the Science and Technology Committee to engage only in science projects. But since our project required the cooperation of many different agencies, the Xiamen Marine Management Office was finally established to play the coordinating roles. Of course, we did not directly participate in the decision-making process of the local government; instead, we only provided guidance and some inspiring ideas for the municipal government to work out the most practical solution for managing such a pilot project. They named the Office themselves and streamlined the functions of the participating agencies. Over the past two decades, we have been on very good working relationship with the Xiamen Municipal Government—we understood each other and things went on smoothly between us. Of course, my ability to communicate in Chinese might have helped in building mutual understanding and trusts.

Despite good communications was maintained, difficulties were encountered. Xiamen Municipality's budget in the 1990s was quite limited, and thus unable to quickly rehabilitate damaged habitats and environmental improvements unlike the US and other developed countries. The UNDP/GEF Regional Program had limited budget for implementing environmental improvements as this was the responsibility of the recipient government. On the other hand, we knew from past experience that even if there was available financial resources, it should not be allocated arbitrarily. Otherwise, it would create dependency, less ownership and sustainability. A project was carried out not because there was enough funding but because the recipient government had the will to implement it. Therefore, we told the incumbent Xiamen Mayor Hong Yongshi, "If you have the will to do the project, we can help you with it. We will work together to move it forward. " He accepted the offer readily.

Frankly speaking, back then, in order to develop its economy, China was carrying out massive structural constructions, sand mining and land reclamation, etc., which was very harmful to the coastal and marine environment. There was no way that we could help the Xiamen government achieved a balance between economic development and environmental protection, but we could raise their awareness of the importance and need to conserve habitats and to protect the environment. We carried out a number of training sessions and activities in Xiamen. For example, we organized scientists to accompany senior municipal officials abroad to see for themselves the effects and consequences of over-development such as large-scale destruction of mangrove forests in some South East Asian countries, which were also well-known for their rich marine resources. Such personal observations and experiences on the sites helped the officials to understand the need for a balanced economic development, a knowledge which cannot be easily acquired through reading books. Before this, the officials in Xiamen rarely had the opportunity to go abroad. However, they were very patriotic and keen to promote economic

▼ During fishing moratorium, law enforcement officials clean up illegal nets.

development in Xiamen, but being aware of the importance of natural environment they were even more determine to protect the environment for future generations. Therefore, it did not take long for the officials to change their mindset in embracing the concept of integrated coastal management in achieving sound economic development without sacrificing the quality of environment. The rehabilitation of Yundang Lake was just one good example of their determinations.

"Coordination" is a buzzword. It is easier said than done. Inter-agency coordination was particularly difficult when we first started. Every agency had its own specific roles and interests and that was why the seas around Xiamen were in a state of "management dilemma" with five different line agencies playing the dominant roles in a limited sea space. With the new institutional arrangement, the Marine Management Office had to play a key coordinating role. With a small group of core technical staff, it was able to reach out to the local academicians and scientists at Xiamen University and the Fujian Institute of Oceanography which provided rich social-economic information and scientific support for management interventions. This approach differs from most other conventional project approach in Xiamen which heavily relied on scientists only. The support of social, economic and natural scientists ensured that management interventions were based on good scientific and economic advice and socio-political acceptance. This was an important evolution from horizontal governance of the past, and increasingly towards science-based management, demonstrating the process of project evolution.

In the past, as also noticed in other countries, most coastal management projects were either driven solely by scientists, the government or NGOs with no or little involvement of the key stakeholders, less collaboration and mutual support. The establishment of the Marine Management Office made it possible for all relevant sectors of the society to work together. This was an important achievement in the first stage of our project.

When we carried out a coastal management project, we must ensure visible results within specific time-frame. Otherwise, the project would have little or no impact and hence, better be left alone. However, recognizing our limited capacity and financial resources, involvement of local government as the driving force is essential. Fortunately the various line agencies of the Xiamen Municipal Government was very enthusiastic and actively involved in project implementation. In fact, before official launching of this project, they had already begun the rehabilitation of Yundang Lagoon. One of the first initiatives of this international project, was to help the government to reorganize the management structure, guide the rehabilitation process and make adjustments/improvements to existing management interventions.

▲ Sea bat farming in Daxie (August 10, 2008) (Photo / Lin Shize)

※ Xiamen model can be applied to other parts of the world for resolving sustainable development challenges.

The priorities of the first phase of the project was to gain confidence in the management approach, strengthen coordinating capacity of the local government, and increase the staff strength and effectiveness of the Marine Management Office. It is important to ensure stronger buy-in by local government and staff so that project activities could be undertaken by themselves. This cultivated mutual trusts between local and international project staff. In fact, the locals were more familiar with their government policies and social system and as such they will be more efficient and effective in executing project activities. On the other hand, in other countries such as the Philippines, public opinion and support from communities are essential elements in ensuring successful implementation of coastal management programs.

Our PEMSEA project started in 1994, while the European Union began their maritime management project in 1995. In March 2017, when I attended a conference in Germany and talked about this issue with delegates from 22 EU member states, I found that their project had long been suspended and they planned to restart it again. They learned that we were able to sustain our ICM practices in the East Asian Seas Region and hence invited me to make a presentation on our best practices. I took the opportunity to introduce the Xiamen Model to them. Over the past two decades, coastal management in Xiamen has evolved from issue-based management to a holistic, integrative management system. Any line agency involved in any aspect of marine and coastal management must make effort to share the information with the relevant authorities or concerned agencies before it is implemented, strongly suggest that an inherent self regulating system has evolved.

Since as early as the 1970s, the issue of sustainable development has gained attention in the international arena. Many international conferences on this topic were organized in the 1980s and 1990s. Now, sustainable development has become the pursuit of almost every country in the world. SDG includes 17 specific goals and is a very complicated agenda of local, national, regional and international implications. The world hosts a sustainable development summit every decade since the 70s, but so far no country dares to say that they have succeeded. But now, we can tell the world that Integrated Coastal Management (ICM) such as the one instituted in Xiamen can be the example of a local solution to the global agenda. In ICM, the key player is the local government, and while outsiders might think that its influence is quite limited. However, once these ICM projects are implemented in increasing cities and coastal areas, it could soon cover a

▲ Xiamen International Ocean Forum was held in 2012. Cai Chengxuan (the second from left) attended the forum as chairman of the board of directors of Partnerships in Environmental Management for the Seas of East Asia (PEMSEA).

substantial length of the coastline. With geographical and functional scaling up of ICM practices, management intervention could cover the entire country or region at large. At present, 14% of the East Asian seas coastline are covered by ICM practices. We are the only region close to achieving the international target of 20%.

Apart from the initial phase, we did not offer any financial assistance for the continuation of the Xiamen ICM program. So how is it possible that its ICM program could continue for over 20 years? The answer is that people have seen the benefits it brings. It would be a lie to say that ICM has not encounter any setbacks in the past 20 years. Changes in tenure of government, conflicts of interest among different agencies and sectors, and many "people-related" problems had to be overcome and resolved. The resolution of these problems largely depends on its value to the people of Xiamen and the dynamism of the Xiamen government. For one thing, the residents of Xiamen have a deep sense of attachment to the marine environment. For another, people have seen the results of management interventions and are naturally motivated to ensure sustainable use of the beautiful island. As governments are more committed and hence the gradual internalization of the ICM practices as part of the regular government program.

▲ The relationship between the ocean and Xiamen is like the relationship between mother and son. The breath, the sound, and the story of the sea are always so lively and vivid. (Photo / Wang Huoyan)

Under Xiamen's ICM, apart from the record-setting revitalization of Yundang Lake and the development of Wuyuan Bay, there have been several impressive achievements:

First is the relocation of roads. I think this is a very important change. The seaside roads of the whole Xiamen Island have been relocated farther inland to turn the coastline into a public resort, where the public can get in touch with Mother Nature. This helps to educate the general public about the importance and value of the adjacent seas.

Second is the change of the function of Gulangyu Island. This was a very wise decision. Many heavily polluting factories on the island have been shut down or moved out of the island, thus playing a key role in protecting Xiamen's seas. Gulangyu Island has been upgraded from a simple residential area to a scenic spot. The new influx of tourists not only brings income to Xiamen, but can also demonstrate and spread the successes of Xiamen's ICM. This is a typical story of "killing two birds with one stone".

Third is the rehabilitation of the wetlands at Wuyuan Bay. Had it been in another city, the government would have used the land for commercial purposes, but in Xiamen, where every piece of land is worth a fortune, the government nevertheless set aside this land to grow mangroves. This also demonstrates that the government is aware that habitats are important natural assets that must be protected and conserved.

Fourth is the exit of cage and oyster farming from the navigational channel. Having once worked as the chairman of the Asian Fisheries Society, I have always cared about the interests of fishermen and fish farmers; but when aquaculture poses environmental damage, blocked sea navigation in the sea channels and affected the city implementation of its functional zoning scheme, I fully endorsed its exit and being transferred to newly designated sites. When Xiamen decided to move out 5000 fish farms from western seas, it has invited immense criticism from abroad. Even I myself was criticized at several international conferences/meetings because of it. However, from the perspective of ICM implementation, the resolution and determination shown by the Xiamen government explains where it differs from many others.

※ Xiamen will produce many more dynamic coastal planners and managers in the near future

I always describe ICM as a process where "one can see the beginning but not the end". When one does something not for personal interests, but out of belief and conviction, one has the courage to continue. Even when one retires, the successors will inherit that belief and continue with the job generation after generation. That is why behind the blue Xiamen seas, we see more and more "sea people."

Of course, the success of Xiamen's ICM hinges upon science, technology and government efforts, but public participation also plays a significant role. Over the past 20 years, I have come to visit Xiamen almost every year, first to follow up project's progress and later for visits after my retirement. I am very pleased to see that every time I come to Xiamen, I realize that the citizens are more conscious of its marine environment and care about preserving its natural beauty..

The Xiamen model is the first success for PEMSEA. This gives us great encouragement, for our idea has therefore become grounded in facts and results, instead of empty talk. Today, when we talk about ICM in international conferences in many cases, we refer to the Xiamen Model.

Although I have already retired 10 years ago, I still follow the progress and impacts of Xiamen's ICM practices. I believe that after over 20 years of practice, the Xiamen model has already contributed to the advancement of a new coastal management science. If academic institutions can further advance this emerging management science into appropriate curriculum, the world would be able to benefit from a new breed of coastal planners and managers to better manage their coastal and marine areas.

For me, competent coastal managers shall possess the following desired qualities:

First, he must think like a scientist. In ICM, science plays an important role, cultivating system thinking, developing science-based management strategy and action plans, implementing risk assessment and management, monitoring progress and scientific reporting, etc. Even when scientific data is temporarily unavailable, they must resort to other technical considerations. "Not enough data" cannot be used as an excuse to say "it cannot be done. "

Second, he must work like a manager. Working as a manager is not easy, as he has to reach out to people from all levels and be good at interpersonal relationships. He should have good communicating and management skills. Only by doing so can he marry scientific ideas with administrative goals and coordinate the relationships among all actors.

Third, he must talk like a diplomat. In ICM, one has to get into contact with people from all walks of life. Be it scholars, farmers, workers, merchants or government officials, one must make sure that he can get his ideas across in a relatively short period of time. For example, within three minutes if he has to communicate with a busy political leader. Only in this way can he reduce the cost of communication. In particular, such skills are essential in soliciting finances and collaboration with local, national or international partners.

Experts with these three traits are hard to find in universities, but I hope to see more of them emerge in Xiamen when ICM has evolved into a new branch of management science.

Export Integrated Coastal Management Experience from Xiamen to the World

Wang Chunsheng

▼

Wang Chunsheng was born in Zini Town, Longhai City, Fujian Province and studied Road and Bridge Engineering at Fuzhou University. In 1983, upon graduation, he went to Xiamen to start his first job at the Municipal Transportation Bureau as a"selected excellent student". In 1996, management of ocean resources was very chaotic and as Xiamen initiated management over marine pollution, Wang was transferred to the Municipal Marine Management Office, responsible for coordinating coastal management. This shift of work allowed him to begin his career in managing and protecting the seas for 20 years and longer.

Wang grew up near the seashore. His home was situated in the middle harbor of the Jiulong River and he set out to sea with his family and villagers when he was little. In the morning when the tide ebbed, a large tidal flat would be revealed and that was the high time to catch shellfish or other seafood. Since he was 15 years old, Wang had begun to sail on a small boat with his brothers towards Xiamen for over an hour, from the middle harbor of the Jiulong River to its river mouth and caught quite a large haul of fish.

In his memory, the Jiulong River was crystal clear and readily drinkable and could be easily collected even from the doorstep of his home. While at sea, even if you could not drink salty sea water, you could always use sea water to clean seafood. However, in the mid-1990s, when Wang was involved in coastal management, things changed both in the seas around Xiamen and the Jiulong River which is upstream. Mud and sand deposited on the sea floor, waste floated on the surface of the oceans, and a stinky smell kept coming ashore. Seeing the beautiful seas in his memory ruined, Wang was saddened and he became determined to make the seas clean and clear again.

Over the next 20 years, Wang experienced every historical juncture in the Integrated Coastal Management of Xiamen, including the Marine Functional Zoning, management of aquaculture, the opening of the Causeways, coordinated coastal use, marine biodiversity conservation as well as the application for the integrated management program of Xiamen Bay (also known as Hap Mun Bay). His experience was also an epitome of the history of Xiamen in coastal management.

Looking back over the last 20 years, Wang would honestly admit that Xiamen is still not as good as he had hoped. But it is still a very beautiful city. It is increasingly true that"the sea is in the city and the city is on the sea". Wang said, "Many say that people in Xiamen lead a very comfortable life and so they have no motivation to work hard. But I don't think so. I believe it's just because people in Xiamen have sea water in their blood and retain some of the habits of fisherfolk. We go fishing every day and no matter how much I get today, I will continue the next day. I can tell many young people in Xiamen are courageous, ambitious and hardworking. If we can provide them with an enabling business environment, they will have a sea of opportunities to start up their own businesses. Though there are rules that need to be followed, red tape can be cut and the world will be their oyster. "

※ From chaotic management to integrated management

Talking about the Integrated Coastal Management in Xiamen, we should begin with the establishment of the Marine Management Office.

In 1996, like many other coastal cities, Xiamen went through a period of chaotic management over the seas with multiple departments managing the seas without coordination or synergy. Departments involved in coastal management included maritime departments, aquatic product departments, shipping channel departments, maritime transportation regulators, the port management department, environmental protection department, marine administrative districts, the Marine Division of the Science and

Introduction to the speaker

Wang Chunsheng, born in Longhai, Fujian, Bachelor of Engineering, senior engineer, is currently the Party Secretary and Director of Xiamen Municipal Bureau of Oceans and Fisheries. Since 1996, he has participated in the establishment of Xiamen Ocean Administration Office, Xiamen Marine Functional Zoning, the establishment of Xiamen Municipal Bureau of Oceans and Fisheries, the construction of the Huandao Road (a ring road lining the island), management of the west sea areas, and Yundang Lake rehabilitation. He is a witness to the 20-year Xiamen Integrated Coastal Management.

Technology Commission, and the Navigation Mark Division, etc. With so many departments doing things on their own, they could accomplish little and the municipal government had to step in and increase coordination. But even the government needed some organs to implement specific work. At first, the Marine Division of the Science and Technology Commission was designated to do this job, but with its limited reach, it failed.

As multiple departments' chaotic management over the seas got worse and worse, the municipal government was faced with many a headache. What headaches? They were mainly caused by conflicts of interest among different industries. For example, when ships sailed in, if an area had been taken up by aquaculture, they simply couldn't possibly move forward. In port construction, aquaculture facilities must be removed. But who should be responsible for this? It was impossible for aquatic product departments to do the whole job. Back at that time, there were also severe conflicts of interest among different departments. So in 1996, the municipal Party Committee and the municipal government decided to set up the Marine Management Office of Xiamen Municipal People's Government as a secondary institute of the General Office of the government, with a goal of coordinating marine development. It was at that time that I was transferred to the Office.

In 2001 when coastal management had progressed to a certain level, the municipal Party Committee and the municipal government decided to merge the Aquatic Products Bureau and the Marine Management Office, and set up the Ocean and Fisheries Bureau of Xiamen. After its founding, there was a focus issue on whether the Marine Management Office should continue to operate. As the Office was leading the integrated management over the western sea areas, a very important aspect of the Integrated Coastal Management, it was decided that the Office should continue to run, but only as a coordination agency. Since then, the Integrated Coastal Management has been conducted on two fronts: administrative regulation by the Oceans and Fisheries Bureau and coordinative management under the Marine Management Office, a very unique mechanism. Such a two-pronged approach represented both industrial regulation and governmental coordination. So the Office continued to exist and worked together with the Oceans and Fisheries Bureau. The Office was headed by the Deputy Secretary General in charge of coastal management of the General Office of the municipal government, while its executive deputy director was the head of Xiamen Municipal Bureau of Oceans and Fisheries. Despite many rounds of reform, this mechanism is still in use today. Looking back over the past 20 years, we can say this mechanism has been a success.

Upon the establishment of this mechanism, Zhu Yayan, then mayor of Xiamen, was appointed head of the Leading Group of Integrated Coastal Management and afterwords successive mayors of this city have also held this post, ensuring its continued authority to foster coordination between departments. Since its founding, a major task of the Office has been port construction in Xiamen. City development relies on construction of ports. Though we met with many challenges in port construction, we have risen up to every one of them with our "problem-oriented" philosophy.

The first task we met was addressing

the problems in dock construction for the Mingda Glass Factory, including the removal of aquaculture in the area, which was also related to cleaning up the waterway. The second task was related to cleaning waste on the sea when Xiamen was trying to build itself into a model clean city. This concerned many departments. Third, we needed to regulate unauthorized vending around Huandao Road. There were still vendors selling products in the old-fashioned way, especially at night. The government had been concerned with this problem for a long time and the task of regulation was assigned to the Office.

There were only 9 people in the Office upon its founding. On founding the Office, we were considering organizing a team based on military standards to supervise coastal management. Therefore, we created an enforcement mechanism and founded the Xiamen Coastal Management Supervision Squad with 18 people who wore uniforms just like city inspectors, but with the name of the Office, and a certificate of administration with an arm badge affixed on top. Our first victory was achieved in the western sea areas by helping Mingda Glass

▼ The Ring Road that lines in the island

Factory to build its dock.

Later on, we met with a series of problems in law enforcement. A lot of aquaculture facilities had to be removed, but who should do the job? Where was the legal basis? Without aquaculture, how would people make a living? We conducted surveys regarding these questions. Technology was not quite as advanced back then and there was no good technological approach available. Therefore, we had to measure one by one by ourselves. It took us years. We turned to several measurement firms and organizations for help, but the risks were too big for anyone to take, so they also turned us down. Eventually we decided to solve the problems of the glass factory first and accumulated some experience for the future.

As for waste on the sea, we also considered organizing a team to deal with it. But on second thought, we decided to coordinate the work and have the Urban Environment Division take charge of it. The Division was under the administration of a municipal office under a different mechanism. In coordinating this work, we also studied on how to get it done in a best way.

※ The Xiamen Marine Expert Panel is founded to render technological support

In building the mechanism of coastal management, we were faced with many problems, which could not possibly be resolved by the Office itself. There were many scientific research institutes in Xiamen. How could they play a part in this process? How could we enable the government to make decisions scientifically? With these questions in mind, the Xiamen Marine Expert Panel was founded.

In the preliminary stage of the Integrated Coastal Management, the expert panel mainly consisted of technocrats, like Zhang Binsheng from the legal field, Zheng Jinmu from the economic field, Lin Hanzong from environmental protection, and Dai Huiwang from port construction. As the preliminary stage was critical, almost all areas needed to be covered, including port, environmental protection, economy, geology, ecology, and aquaculture. We knew several institutes, colleges and universities engaged in marine research, so we approached them to seek the right researchers to be members of the panel. Finally, we gathered together technocrats and researchers that almost covered every field concerning oceans.

After the first batch of experts retired and the Oceans and Fisheries Bureau was founded, most members of the expert panel were researchers. This was the secondary stage. In this stage, we tried to separate decision-making and technological support so as to exercise power independently and keep technological support free from the influence of government officials. Most of the researchers came from Xiamen University, Jimei University, the Third

1 | 2

1. The Law Enforcement of Xiamen Seas vigorously improves activities in the seas of Xiamen.
2. The Law Enforcement Team of Xiamen Seas

Institute of Oceanography of the State Oceanic Administration, Xiamen Institute of Design, Xiamen Planning Institute and some other enterprises, basically covering all areas related to oceans. Furthermore, by then a legal framework on coastal management had basically come into shape, so the legal part was no longer a focus. What"s more, as the concept of scientific governmental decision-making was promoted, intervention by officials was also reduced.

Members of the expert panel mainly fall into two categories. The first is experts in economics, most of which are national-level experts, including 7 or 8 academicians. As marine economy developed to a certain level, we were more than eager to get technological support from experts in this area. The remainder are experts from other areas.

Since 2016, more and more young members have joined the expert panel because now there are quite a lot of people from government agencies who understand both technology and management. So now what we are considering is whether we need to also have experts in economics and law on board. After careful consideration, we have decided that we do need them. We are working on integrated management, so experts from only one field cannot meet current demands and we need to put expertise from different areas together.

As organizer of the expert panel, the

1
___
2
___
3

1. "The Comprehensive Rehabilitation Plan for Ban on Aquafarming in the West Seas" solicited opinions from the National People's Congress.
2. "The Comprehensive Rehabilitation Plan for Ban on Aquafarming in the West Seas" soliciting opinions from the Chinese People's Political Consultative Conference
3. Follow–up media reports on the progress of Xiamen's west seas rehabilitation.

Office only provides service to the group and has no right to administer it. Instead, it is under direct management of the municipal government and is also funded by the government. In the preliminary stage of the integrated management, much work was assigned to the panel, which acted with a sense of justice and avoided conflicts of interest. After the Oceans and Fisheries Bureau was established, the Office becomes a coordinator and basically takes charge of no specific affairs. Instead, the Office assigns tasks to specific organizations through the expert panel and it only provides suggestions for the final reviews of these tasks for reference. So what will we do on the next stage? We are still thinking about how we position ourselves and our direction for the future. There will be many changes under the 13th Five-Year plan. So things have yet to be settled. But now, our target is to optimize technological support for coastal management.

Though experts in the panel are individuals, they represent a series of research institutes, colleges and universities. What we are considering is how to give full play to them and gain better support for our work. It will be an issue worth thinking over in the years to come.

Normally, there are 15 members in the expert panel. When retired, in principle, heads of the panel are appointed counselors. There is no fixed amount regarding research funds assigned to the panel from the government. But the funds are quite substantial when compared to our counterparts all over China, and funding is increasing year on year.

※ Marine functional zoning opens up a new chapter for coastal development in Xiamen

After the founding of the expert panel, the Office has assigned several projects to it. The most important one concerns marine planning. Urban planning does not cover marine planning, so the panel needed to formulate a marine functional zoning scheme.

At first, we only suggested some ideas to the panel on whether we could divide sea areas in Xiamen and set up several functional zones identified by their dominant functions, compatible functions, and restricted functions. By then, there happened to be an international cooperation project about to be implemented in Xiamen by an organization called Partnerships in Environmental Management for the Seas of East Asia (PEMSEA), cofounded by UNDP, IMO and GEF. With funds from that project, we mobilized relevant organizations in Xiamen to provide the relevant materials for the expert panel to draw up a scheme on marine functional zoning. Eventually, we designated 4 functional zones for Xiamen's marine areas, namely western sea areas, eastern sea areas, Dadeng sea areas, and Tong'an sea areas, and each zone had clear restrictions on aquaculture, which provided a legal basis for the withdrawal of aquaculture.

Looking back, we can see that our work at that time was consistent with the plan for dominant marine functional zoning published later by the country.

After publication of the scheme on marine functional zoning, we had problems concerning its implementation. So we immediately got to work introducing *Provisions on the Use of Sea Areas of Xiamen*, and with this legal basis, our plan became legally enforceable. This was also an important part of the decision-making process.

However, with regards to how to coordinate marine functional zoning, land planning, port planning, and environmental protection, there were a lot of disagreements among the different departments. Later on, we decided to leave this problem to experts and then to the municipal Party Committee and the municipal government to decide. This process ensured that our decisions were scientific and in this process, experts contributed technological support to the decision-making process.

Though there are many daunting challenges on the way ahead, we are fearless and ready to rise up to them.

For example, we proposed several plans for the management of the western sea areas. But if we implemented these plans, we would run into big trouble. When we developed the economy, we often reclaimed sea areas for land use. But small as it is, Xiamen had very limited sea areas for reclamation and we must set rules on it, otherwise the sea areas would disappear. Therefore, apart from making plans, we also needed to resolve technical issues on what to do and how to do it. So experts studied the sensitive areas vulnerable to reclamation. When we built the Huandao Road, we conducted a research study over the eastern sea areas, concerning where to draw the red line to control reclamation and how to safely discharge pollutants.

Actually, we exercised the most restrictions in the western sea areas, including Maluan Bay and Jingkou Bay, the now Dongyu Bay. There was not much land in Haicang Bay and Dongyu Bay was a shoal historically. So people thought this area should be reclaimed. After research and investigations, we set three restriction lines and the last one was settled legally. You can still find this boundary line in *Provisions on Marine Environmental Protection of Xiamen*.

As the dominant function of the western sea areas is ports shipping, we must ensure that the waters remain largely free from siltation. To achieve this target, we had to open up the Gaoji Causeway and then Maluan Bay Causeway. Before the No.14 Typhoon in 1999, we had discussed the possibility of opening up the Gaoji Causeway. After hydrodynamic calculation, we believed it was plausible. But this decision might not be approved by people and it would be very expensive. So we put it aside and discussed instead opening up the Maluan Bay Causeway first.

This was one of the solutions we proposed after research and investigations. The purpose of integrated management over the western sea areas is to protect the Port of Xiamen and ensure several key factors are upheld. One of them is for reclaiming Haicang Bay, we must open up Maluan Bay and only by opening up Gaiji Causeway can we solve the problems related to Haicang Bay. Only after all this work is done, can we initiate integrated management over the western sea areas.

According to the Marine Functional Zoning, coastal tourism is the dominant function of the east coastal areas. In building the Huandao Road, we also met the issue of the restriction line: Where was the line to be put, inward or outward, within the beach or outside of the beach? Basically, our principle was that first, we needed to protect sand beaches, second, to ensure smooth passage for main navigation, and third, to maintain the beautiful environment in Xiamen. After rounds of discussions, we decided to keep the road from Xiamen University to Xiamen Conference and Exhibition Center and if necessary, we would build bridges in some regions instead of reclaiming sea areas.

When building bridges, there still remained a big issue. As the port for the Sea Rescue Service was located where we planned to build a bridge. Should we remove the port or reclaim the navigation channel? Eventually, we decided to remove the port.

※ Legislation resolves issues of space and ecological problems

When discussing marine legislation in Xiamen, we should begin by discussing legislation over the Huandao Road.

Huandao Road was originally a prohibited area for war preparation and opened to public later on. Then, some people secretly excavated sand from there, which damaged the environment and the beautiful coastal scenery. Who should exercise regulation over such affairs? Nobody knew. So later on, *Regulations on Managing the Resources of Sands, Rocks and Soils* was introduced, and sand beaches and coastal lines became legally protected. To put it in terms of a popular term today, we would be "problem-oriented".

Back then, there was an overall plan for the development of Xiamen and one part of it was

▲ Awareness campaign before rehabilitation

about social and economic development. So based on the plan, we made clear the dominant functions, compatible functions and restricted functions of every functional zone. What was to be done after that? Who should be the one to ensure the proper implementation of the plan? To answer these questions, Xiamen adopted Provisions on the *Use of Sea Areas of Xiamen*, founded the Marine Management Office to take charge of integrated coordination and identified marine functional zoning as an important plan compatible with the overall development plan of Xiamen. All relevant project construction, including projects on ports, aquaculture and environmental protection, was required to conform to the plan. This exemplifies how legislation protects marine space.

With this done, we met with another new problem: protection of the Chinese white dolphin. At that time, the Chinese egret and lancelet were among China's second-class protected animal species, but not the Chinese white dolphin. To protect the Chinese white dolphin and ensure biodiversity, Xiamen formulated Provisions on the Protection of Chinese

White Dolphins of Xiamen. This is a case showing how legislation protects biodiversity.

Back then, there were three pillar industries in Xiamen: ports and shipping, tourism, and aquaculture. Aquaculture alone occupied over 120 km², one third of the total area of Xiamen at that time. But these industries had severe conflicts and troubles with each other every day. For example, aquaculture blocked navigation channels and also affected the scenery for tourists. The founding of the Marine Management Office aimed at resolving these issues. Based on this, Xiamen also introduced *Administrative Rules for Aquaculture in Shallow Seas and Tidal Flats* and later on *Regulations on Port Management* was also adopted to tackle problems concerning ports. So it seemed the issue of space and industrial conflicts could also be resolved legally.

▲ Xiamen introduced *Provisions on Marine Environmental Protection of Xiamen* to provide a legal basis for environmental conservation.

On the next stage, legislation was designed to resolve pollution, including pollution from ships, aquaculture pollution, pollution from road shoulders, as well as basin pollution. Back then, the most serious pollution in Xiamen was in the Jiulong River. To tackle pollution, Xiamen introduced *Provisions on Marine Environmental Protection of Xiamen* to provide a legal basis for environmental conservation.

Based on real demands, we adopted over a dozen of regulations on coastal management to tackle emergent issues whenever they popped up. Through legislation, we successfully established a legal framework for coastal management and acted in its accordance, with the Marine Management Office as the coordinator. Over the years, we have accumulated much experience and continued making amendments to them in light of relevant national laws and regulations.

How did these regulations come into being?

After the Office drafted the regulations, we looked for questionable points in the drafts according to our experience and made modifications based on the national legal framework. We then submitted them to the Bureau of Laws for experts to review.

In the initial stage of legislation, after the draft was produced, suggestions were solicited from different sides. As each department considered from their own perspective, it was very difficult to reach consensus and finalize regulations. In legislative conferences, it was very usual to have heated arguments. Later on, we learned to make compromise. We would pass items we agreed on and put aside those for which we did not agree. After rounds and rounds of negotiations, when we more or less reached a consensus, we would submit our draft to the Bureau of Laws to review from a legal perspective. After the review, we would submit them to regular meetings of the government for deliberation and then introduce the regulation if passed.

▲ The first Chinese white dolphin reservation base is located in Huoshao Island, Xiamen.

Introducing legislation is a very complicated process, especially in internal discussions where everyone has a point to make. But we were patient enough to carry forward rounds of discussions. For example, recently we amended *Provisions on Marine Environmental Protection of Xiamen*, which took over 20 rounds of revisions before reaching internal consensus.

In terms of legislation over marine issues, Xiamen led the whole country, and to some extent, affected national marine legislation.

In setting the red-lines of control over the eastern sea areas in Xiamen, we did a lot of research and decided to incorporate Wuyuan Bay, the airport and the Huandao Road within the last line. In later legislation, we protected the coastline of Haicang Bay and forbade its development. After *Provisions on Marine Environmental Protection of Xiamen* was amended in the 1990s, it provided that several coastal lines were to be protected. Now, at a national level, the regulation of natural coastlines has fixed standards. It would be fair to say that our practice guided national regulations and our local experience contributed to the efforts of the central government.

Before *Law of the People's Republic of China on the Administration of Sea Areas* was published, I stayed in Beijing for over half a year to do just one thing every day: work together with the Law Committee of the National People's Congress that was responsible for the legislation. They had many theories but little experience. So I told them about the legislative process for implementing relevant regulations on coastal management in Xiamen, including including the relationships between the regulations. In fact, *Law of the People's Republic of China on the Administration of Sea Areas* was formulated based on Xiamen's legal system of coastal management.

※ The 4 ways to protect sand beaches within the Huandao Road

As we carry forward our work, we are conscious of the need to protect our sand beaches. But the question is how? A typical example is the construction at the gate of the Third Institute of Oceanography of the State Oceanic Administration. We usually build an upright wall during construction, a simple and effective method. Staff from construction department generally believe it would be enough just to leave it there and don't give the sand beaches much attention. As I majored in road and bridge construction, I am quite sensitive to landscaping. I wondered whether we could have another solution. Through much effort, we built an arc-shaped wall there to resist waves and preserve the beaches, as you can see today.

The guiding principle behind the construction of the Huandao Road was to "Staying close to the sea, and endowing the beautiful beaches to our people". When building road sections, they were either constructed near mountains or seas, and bridges were built in coastal areas, or were cut into a curved shape, or were tunneled through rock and caves. Their construction was based on high and strict standards, which displayed our respect for the nature and the seas.

First, we built the Yanwu Bridge near the first beach along the Road. This is probably the bridge lowest to sea surface in the world. However, construction of the bridge would damage the environment if it were based on traditional building standards. So we had to break away from tradition. But, what standards should we use then, the municipal construction standards or the highway standards? This question was hotly debated. After a bitter process of negotiation, we decided to combine highway standards with municipal standards as well as add our own designs to establish a suitable standard for construction of the bridge with the goal of protecting the coastal lines of beaches. In this way, we met both the technical requirements and landscaping demands. We made special designs to minimize the deleterious effects on beaches. Even the bridge floor was elaborately designed.

Second, we conducted programs to restore beaches to their original state as we built roads along the beaches. When the No. 14 Typhoon hit in 1999, the typhoon was so powerful that it blew away our patrol trucks. People had to cover their heads and put their faces to the ground on all fours to keep themselves from being blown away. After the typhoon, all buildings on the sand beaches were blown down and many coastlines were damaged. In contrast, those buildings and facilities that were built outside of beaches, out of respect for nature, stayed intact.

It is impossible for men to conquer nature. Learning from this experience and in case of similar accidents in the future, we raised over 400,000 RMB for the protection of beaches. We planted coconut trees and morning glories in the beaches and built a windbreaker on shore to create shade and stabilize the sands, which also contributed to landscaping.

Third, from the Conference and Exhibition Center to Xiangshan, and to Wutong, there was a very well-preserved coastline, but it was not easy to access. So we left quite a large space clear close to the sea and built ramps and auxiliary paths to allow people to enjoy the sea. It is quite comfortable to walk on the auxiliary paths, breathing in the salty ocean air.

Additionally, the area from Xiangshan to Wuyuan Bay (then Zhongzhai Bay) was enclosed by fishermen for aquaculture. Regarding enclosing the sea area for cultivation, there were two opposite opinions among experts and the government organs. Many experts believed highly-saline earth and wetlands should all be reclaimed for development because every inch in Xiamen was valuable. While as government agencies were responsible for the oceans, we hoped to clear out salt factories, build bridges, dredge the seafloor and conduct hydraulic fills and reclaim land for seas so as to increase coastal lines and expand sea areas.

The experts themselves didn't see eye to eye with each other. Some objected to building bridges and supported land reclamation, because it costed too much to build bridges while reclamation could help reduce wind erosion towards Xiamen Island. However, experts who advocated for building bridges but objected to reclamation believed erosion was not a big problem and we could turn to other ways to tackle erosion, like planting mangroves. To reach consensus, we invited experts and scholars from Nanjing Hydraulic Research Institute, Xiamen University and the Third Institute of Oceanography of the State Oceanic

▼ Shore near the Convention and Exhibition Center before and after the rehabilitation

Administration, and utilized many mathematical and physical models to review these opinions based on different working conditions and targets for environmental indicators. Eventually, it was agreed that if highly-saline land and wetland areas were reclaimed, there would be only 300 to 400 meters of coastline in Xiamen. We would have more land, but only the side facing the sea was valuable. However, if the land was dug out for bridges, there would be an additional 8 km of coastline in Xiamen, all very valuable. So during a conference of the municipal standing committee, after listening to opinions from both sides, mayor Zhu Yayan decided to build bridges.

There was a detail worth mentioning. There were two large rocks in Wuyuan Bay, one of which is the "Sailing Rock". Geographically, these two rocks hardly affected navigation channels. But as there were very few rocks in the seas around Xiamen, we would do our best to keep them. Besides, from the perspective of biodiversity, rocks serve as a unique habitat for marine species in the area. This was also true in the construction of Wuyuan Bay. It was standard practice to dig 7 meters under sea and we told construction personnel to dig more holes and dig deeper in some areas, filling them in with stones in order to boost biodiversity.

After the wave-resistant walls were completed, we started to deal with the tidal flat nearby. It was quite an eye sore and some people suggested planting mangroves there. But there were three problems with this suggestion. First, mangroves could cause more sedimentation; second, there was no variety of mangroves suitable for Xiamen; third, maintenance cost for the mangroves was expected to be very high. As historically, this area was a sand beach, we decided to restore it back to its natural state. After all this time the sands are still there. Now, we are working on linking the beaches from Wutong, to Guanyin Mountain and to the Conference and Exhibition Center, in efforts to improve the

▲ The beach design and planning of Xiamen served as a model for coastal cities of the country, which was introduced to the United States.
(photo / Wang Huoyan)

environment of the whole region. The design and planning of sand beaches in Xiamen served as an example for the rest of the country and now, Jiangxi Province, Qinhuangdao, Pingtan are all learning from us. This experience has also been studied in the United States.

So, how can building beaches contribute to protecting the marine environment? First, beaches serve as infrastructure for coastal tourism and also help protect natural coastlines. Besides, beaches help to improve the local ecology. But now, due to too much human intervention, a problem has emerged. Namely, once the beaches are completed, they become a comfortable habitat for short-necked clams. Then, when people come to beaches to catch short-necked clams, they disturb the habitat of lancelets. Now although we are making great efforts in monitoring and regulation, it is a difficult problem to solve and we must enhance publicity and education in this regard.

Additionally, beaches are buffer to powerful waves. In the past, when typhoons came, waves directly slapped against the shore, while now, sand beaches can prevent waves from coming ashore. Mangroves can also play the same role. But mangroves have specific habitat requirements and we make plans accordingly. Thereis also a small sand beach in Pearl Bay, but its berm is too short to retain sand. So when waves come, the sand erodes away.

Why are mangrove forests under the purview of marine agencies in Xiamen? Some have suggested that mangroves should be under the administration of the forestry department. It is a long story. When we started coastal management, we didn't know how to delimit sea areas. With the zero-meter line or the high-water line? It was hotly debated. Eventually, we settled on the standard of average high-water line for large waves. However, according to land department standards, it should be the zero-meter line and tidal flats should also be included. The forestry department also held that 6 meters above sea level was categorized as tidal flats, so those areas should be put under their administration. After several rounds of arguments, we reached no consensus. This was a difficulty we encountered during legislation and another emerged after legislation: how to resolve the conflict between laws and management. Which should we follow, existing laws or local practices? The question of the management of mangroves is a typical example. I proposed in the Marine Management Office that what proves to be effective is correct. So we prepared special plans for the conservation of wetlands and implemented our plans step by step. We started trial planting of mangroves in Haicang Bay and then in Xinglin Bay, and then Yundang Lake. At first, people doubted whether mangroves could survive. But ultimately, we succeeded in our trials. After the introduction of Law of the People's Republic of China on the Administration of the Use of Sea Areas, when we reported our work to the municipal Party Committee, we listed all we did, including the management of mangroves forests. However, the forestry department still insisted mangroves should be under their administration. The municipal party committee ruled: first, the Marine Management Office had accomplished quite a lot in mangroves management; second, coastal management was within the scope of the duties of marine departments like the Office; third, based on *Provisions on the Use of Sea Areas of Xiamen*, mangroves along the seaside should be managed by marine departments. After years of practice, mangroves are still managed by marine departments in Xiamen, an arrangement unique in the country. Though, the forestry department has already been merged with the Municipal Bureau of Parks, it is still well acknowledged that mangrove wetland parks are under administration of marine departments.

What are the benefits of marine departments managing mangrove forests? It is beneficial for us to incorporate mangroves into the whole picture of marine environmental rehabilitation, taking all elements together into account, including utilization of tidal flats. Besides, we also combine mangrove management with marine economic development, like with tourism, to create further dividends. What's more, we also consider mangroves part of the protection and conservation of the marine ecological system and biodiversity.

Today, beach construction is complete on Gulangyu Island and essentially across all of Xiamen Island. We only have the last 1.5 km to go, namely, the part from Guanyin Mountain to Wutong, which is still under construction. Across the whole Huandao Road, only the part from Battery of Huli Mountain to Music Square is not suitable for sand beaches and we have to use pebbles instead. But in summer, pebbles are too hot and difficult for women to walk on with high-heels, so at first, there were many complaints. But people got used to it gradually. In the past, the area had to be repaired or torn down to rebuild every year because of typhoons. Now, after building pebble beaches, this problem has been resolved, as the beaches absorb water.

Now, sand siltation carried from the waters of the Jiulong River is decreasing. Sand dams are an important mechanism to collect sand and return it to beaches. There is a very good sand dam on Xiajin Water Channel between of the exit of Tong'an Bay and Great Kinmen and Lesser Kinmen. We consult the map of Xiamen from 1937, and wherever there were sand areas, we study the feasibility of restoring them with scientific tools like mathematical models. If it is feasible technologically, we try to rebuild it. As there is no sand any more, its tides and ecology are bound to be different. But its intrinsic nature remains unchanged. If one place held sand, it might still be able to hold sand. Based on this idea, we have further studied siltation problems, and as mentioned earlier, we have built some facilities to keep sand from eroding.

※ Easing pollution with the "Farmland to Sea" Campaign

After we determined the 4 ways to protect sand beaches along Huandao Road, we started to remove aquaculture.

Since 1998, more and more tourists have come to the eastern sea areas. However, there used to be many aquaculture facilities in the areas. We considered using this region as a pilot for the withdrawal of aquaculture. But there were huge disagreements among experts. Environmental experts believed that aquaculture should be cleared out, yet from the point of view of economic development, marine experts held that aquaculture should be kept and protected. Therefore, after a long tug of debate, we established two aquaculture areas in Tong'an Bay and Dadeng Bay.

In fact, aquaculture itself hardly affects the marine environment. However, because of human intervention, it can cause much influence. The largest problem is that aquaculture blocks navigation channels. Furthermore, it slows down water exchange. Additionally, it causes pollution. For example, when oyster shells sink down under water, it increases siltation. Besides, fish cannot possibly be healthy in such an environment, and are unsuitable for human consumption. At that time, pollution caused by aquaculture was still controllable, so we were determined to address it by clearing out aquaculture and helping fishermen seek other jobs, such as working in the tourism industry. At the same time, we also created a mechanism for transferring the responsibility of removing aquaculture to each district, funded by both the municipal government and districts themselves. The first pilot field was in Lianqian Street of Siming district. Today, compensation for land requisition and demolition still follow the example of this pilot program.

After we solved old problems, new ones popped up.

Immediately after the the withdrawal of aquaculture, many vendors came to the beaches with umbrellas over their stands. After discussion with the expert panel, we all agreed to keep the beaches open to the public, but free from vendors. Sand beaches not only belong to people in Xiamen, but also to people around China. We don't allow anyone to do business on beaches that belong to all. It turned out that our efforts paid off and this phenomenon was curbed.

Practical experience teaches valuable lessons. One regret we had in the management of Huandao Road was that after it was completed, we allowed the use of pavements near the seaside for private enterprise, resulting in Yefengzhai and Huangyepo, two commercial scenic spots. Their operation often blocks the pavement, and it is very difficult to tackle this issue now. So this is a serious problem yet to be resolved.

After the removal of aquaculture, fishermen benefitted. For example, in the past, Zengcuo'an was just a small location, but now it has expanded and become a cultural village. Because of industrial and technological advances, people will be able to copy this experience in the future. If you tell fishermen in these villages to go back to aquaculture, they wouldn't want to and they also wouldn't want to destroy the beautiful marine environment which we have taken great efforts to restore.

Some illegal aquatic commercial activities still remain. Though few in number, they are still there, especially in Qiongtou village and Xinglin village. They received compensation for halted aquaculture long time ago, yet they refused to stop, and so the problem has dragged on till now. What can we do about it? We once proposed making the area a program subject, but it was too difficult and expensive. So there has been no progress in this problem. We have made sure that no additional areas for aquaculture will be added. This is a problem yet

1. After the comprehensive rehabilitation of aquafarming in seas started, fishermen dismantled fishing rafts voluntarily after rounds of consultation and coordination.
2. Over 150 mu of sea bats raising cages located in the southeastern waters of Baozhu Island is the last aquafarming facility that's removed.

to be resolved.

To implement Integrated Coastal Management, Xiamen focused our efforts on implementing a series of relevant regulations, including *Provisions on the Use of Sea Areas of Xiamen, Provisions on Marine Environmental Protection of Xiamen*, *Regulations on Managing the Resources of Sands, Rocks and Soils, Management Measures on Nature Conserves of Lancelet*, and *Regulations on the Protection of Chinese White Dolphin*. Through experience, we have come to understand that after we are equipped with adequate mechanisms, laws and regulations, then it is all about law enforcement. At first, the Office founded the Xiamen Coastal Management Supervision Squad. After the Aquatic Products Bureau and Marine Management Office merged, an Integrated Marine Enforcement Squad was founded, responsible for marine legal enforcement. This is the first of its kind in the whole country. In the first few years, this team played a critical role in the management of western and eastern sea areas and it is still very important today.

While limiting coastal land reclamation in the western sea areas, there remained the question how to realize compensated use of sea areas. The solution lied in the introduction of relevant mechanisms. As early as 1998, Xiamen adopted *Regulations on the Exploitation of Marine Resources for Profit*. Before relevant national laws and regulations were published, we set out a scheme on Marine functional zoning in Xiamen, identifying dominant functions, restricted functions and compatible functions. We collect fewer fees for functions we encourage and have higher fees for industries and behaviors we want to restrict. This regulation was in effect until the adoption of the *National Regulation on Use of Marine Areas* in 2007. It is fair to say we were quite ahead of the curve in capitalizing on economic and legal tools to regulate the use of the seas and implement compensated use of marine resources.

To my delight, after a decade of efforts in integrated coastal management, areas reclaimed for land use and areas returned to the seas are about the same size, with land returned to seas slightly higher. This has much to do with our change in mindset, especially in terms of environmental protection. It is indeed a very important achievement, given the limited land

West and East Coastal Areas

Historically, the area of the west seas was 108 km². The area has shrunk to only about 46 km² after long time blind cofferdam, which has greatly reduced the tidal volume. In 2002, the Xiamen Municipal Government started withdrawing aquafarming with compensation in the West Seas. The rehabilitated area reached 70 km², with total funds 230 million yuan, and more than 20,000 mu of water was free from aquafarming.

The sea area around the East China Sea is about 91 km². Due to large—scale farming, siltation deposited in an area that's almost 50 km², occupying 56% of the sea area. In 2006, through systematic and scientific study, aquafarming was removed from an area of 77.2 km² and 24.5 km² land was turned back into sea. After rehabilitation, the tidal volume increased by 44 million m³, and the water exchange capacity improved by 30%.

areas in Xiamen and their high prices.

※ Opening up a few causeways to unleash vigor

In managing the seas, we also pay attention to public opinion.

People had much affection for the Gaoji Causeway and when we tried to open it up, many people expressed objections. Now things are getting ready and it has been settled to open it up, but such sentiments will not be diminished. For this reason, we plan to build a park and a memorial hall for the Causeway. This affection for the Causeway is indeed marvelous and we are going to embrace it and encourage public appreciation for engineering works.

If ideas are too ahead of their time, we might not receive public support, and as a result our efforts may fail. This was the case during the management of the western sea areas. While marine experts supported our efforts, the majority of the public did not understand us or agree with us, so we met with many obstacles in our work. Since 1998, we have done a lot of work in developing public understanding and support. A turning point came up in 1999 when the No.14 Typhoon wreaked havoc on almost everything in the western sea areas. We were faced with the choice of reconstruction or demolition. Yu Weiguo, the official directly in charge of our work and the now provincial governor, said that we should work towards recovery, restore production, yet also curb the increase of aquaculture. If we removed all aquaculture, we would strike a double blow to aquatic farmers in the area, creating unrest. So we set forward our plans and worked to persuade people of validity of the changes at the same time. In 2000, a remote sensing map was finished and with that, we conducted removal and monitoring activities in that area while striving to ensure justice, fairness and openness.

This is what happened later. Now, let's return to the story of causeways.

Maluan Bay also went through several stages of planning. As people came to accept our plans, the final one was put in place and its result was better than expected. It is estimated that the holding capacity of the bay would be increased by around 30 million cubic meters. After opening the Causeway, we had basically achieved our intended targets. Not only has water exchange improved, but the interchange points of ebb and flow also have changed. However, one question remains to be answered. That is after the Causeway was opened, should we remove sand from the western sea areas or bring in fresh sea water from the eastern sea areas?

Now, we are opening the Dongkeng Bay Causeway. In the management of the western sea

▲ In the 20th century, the mountains were reclaimed and causeways were built to connect Xiamen Island with the mainland. (photo / Li Kaicong)

areas, we only took 4 bays into consideration: Maluan Bay, Wuyuan Bay, Dongkeng Bay and Xinglin Bay while the eastern coast surrounding areas were not taken into account. Because through research and investigations, we found out that if we planned to reclaim land along the west coastlines of Tong'an Bay, we must open up Bingzhou Island. We didn't uphold coastal reclamation on the west bank of the eastern coast surrounding areas. Later on, when the municipal Party Committee and the municipal government led management of the eastern coast surrounding areas, we opened up the causeway and restored Bingzhou Island back to its original state.

※ Managing public relations with wits and courage

With years of experience in Integrated Coastal Management, I learned that it requires not only courage, but also wits to resolve issues with the public.

In early 2002, the Standing Committee of the municipal government decided to submit the plan of integrated management over the western sea areas. When we were discussing

▲ On July 18th, 2012, the high causeway was opened. At 0:40 a.m., the seawater in the east and west waters was being circulated.

whether to manage the areas or not, the then executive deputy mayor Zhang Changping also realized we might have no opportunity to manage the western sea areas in the future if we do not do so right now. Therefore, he instructed us to make a plan and seek suggestions and opinions from every village Party branch. At first, fishermen were all against the proposals, but then they agreed and started to talk about compensation issues. The Municipal People's Congress and the Municipal Political Consultative Committee also raised objections. But after we reported our plan to the then municipal Party secretary Hong Yongshi, he decided to go ahead and implement it. Then, what we needed was a way to unite the people together.

On April 23rd, 2002, a meeting for the implementation of Integrated Coastal Management and the prohibition of aquaculture in the western sea areas was held. All groups of the municipal Party committee and the municipal government agreed to support this work. That was the largest meeting on implementation in Xiamen since I arrived. It showed the determination of the municipal Party committee and the municipal government and also the difficulty of the task ahead.

At the pep rally, it was decided that before Oct 31st, 2002, aquaculture facilities and activities would be cleared out. This was a daunting task and all districts and government organs directly under the municipal government set up working groups and sent officials to the front-line. When we negotiated with villagers, the biggest headaches came from non-local aquatic farmers, and they were also the ones to create troubles later on, because this plan directly hurt their interests. As we conducted our work, the remote sensing map was of great help. With a resolution accurate to 0.6 meters, the map clearly marked longitude and latitude coordinates. Every working group received a copy and with this map, they could easily and clearly check with aquatic farmers about their aquatic areas.

What about existing aquatic industry? The municipal government mobilized officials to buy the industries out. I was also worked with two companies that processed raw fish. It took them about five months to process all existing raw fish stocks.

Setting up standards for proper compensation was also a headache. If the municipal government set the standards, it would lead to a series of problems in later years. For example, with the improvement of living standards, prices for land and sea areas would also increase, as would compensation. So after discussions, we decided that the Marine Management Office should take the lead and 4 districts of the city jointly set up the standards through consultation and negotiation. We worked out compensation standards for net cages, as well as compensations for oyster breeding areas and short-necked clam breeding areas per mu. Then we went to solicit opinions in each district and after reaching consensus, we organized a joint conference of the 4 districts to settle compensation standards. Therefore, we set the compensation standards through democratic consultation, and though not legally binding, it was effective.

So the next question was where would the net cages go once they were removed? Hearing that we had removed net cages, the municipal government of Ningde sent representatives to talk us into helping transfer the net cages to Ningde. But we refused immediately because we knew once the cages were transferred, they would also cause pollution in Ningde. Then they would also come to us when they had to deal with the resulting pollution. This was a very practical issue, so we told the government of Ningde that it was up to those engaged in aquaculture whether they'd like to move to Ningde or not, but we would not encourage them or intervene. Later on, many people did go to Ningde for aquaculture. But what we were worried about was that people might go to Dadeng sea areas and Tong'an Bay instead. So we registered all net cages in these two areas and fixed the total number. Once new ones were installed, old cages would be removed. Did we miss any cages? Yes, I'm sure we did.

But that has become a problem left to be resolved.

As we gradually pushed forward our work, we came to realize that it was not enough to simply remove aquaculture. We also needed development, or we could hardly prevent aquaculture from coming back. So is it possible to simultaneously push forward the exit of aquaculture and promote development of the western sea areas? We began by dredging around the Xiamen Bridge and as the area was not suitable for aquaculture, farmers never came back. As this pilot program was a success, this method became popular.

With the sludge cleared out from the Xiamen Bridge, we built the Tan Kah Kee Memorial Museum. We promoted this experience, leading to the construction of Yuanboyuan Expo Garden in Xinglin Bay. The Water Conservatory Department believed there was heavy siltation in Xinglin Bay and that it needed to be dredged. But where to put the sludge? We first planned to put the sludge into aquaculture pools, and fill them up with just one ditch left for flood discharge. But if so, we needed to get rid of the fish in the pools first, and the process would not be good for future development and would damage the ecology and environment. Besides, in the event of flood, the catchment areas could be as large as over 200 km^2, with only one ditch, the whole of Jimei University would be submerged.

At last, we planned to build 9 islands, which would not only help with flood mitigation and flood discharge, but could also complement the construction of Yuanboyuan. There was the largest wetland in Xiamen, home to many species of birds. To provide a shelter for the birds, we built two bird islands, which could also be regarded as ecological islands. Some doubted the necessity of building two ecological islands, after all Yuanboyuan was built with plenty of trees. But we believed that was insufficient. Many people would visit Yuanboyuan, while the two bird islands could be completely isolated from human disruption.

There were many similar problems in the construction of Yuanboyuan. With the goal of protecting the whole ecological system, the process was not easy. For example, some believed it would be beneficial in terms of the exchange of freshwater and sea water to promote reclamation. But if so, it would seriously affect the ecology of the whole ecosystem.

※ Coastal management through rigorous debate and science-based practice

In the coastal management in Xiamen, rigorous science-based practice is exercised in every detail.

After our management in the western sea areas were under way, aquaculture control over Tong'an Bay and Dadeng sea areas began. We had disagreements within the government. For example, fishery regulators believed that in aquaculture areas, we should encourage aquaculture, and all we needed was regulation. This was in fact how we chose to proceed. If aquaculture affected scenery, we would make efforts to improve the scenery. However, despite such attempts, our scenic facilities were often ruined by surging waves. So our efforts here failed. Later on, we began to require registration in every district. But the authority of handing license lied with each district, so when we went to districts to give them instructions, we found that some fishing companies had more licenses than they deserved as a result of their pulling strings. So we had to enhance administration. First, we needed to ensure proper registration. Second, we needed to enhance supervision and publish notices on banning the expansion of aquaculture. The Dadeng sea area allowed for aquaculture and administration over this area should limit aquaculture within the boundary of the area. As this region was not fit for cage aquaculture, but suitable for raising oysters

▲ In June 2002, subsidies for dismantling cages in Haicang area were given out for the first time.

1. On June 10, 2003, the inspection meeting for withdrawing aquafarming in West Seas was held.
2. Large breeder in the west seas (first from right)
3. Zhang Changping, then deputy secretary of the CPC Xiamen Municipal Committee and standing deputy mayor, met with the village party committees to get information on the ground.

by putting them on lines, supervision of the boundaries was possible.

Currently, the sea reclamation in Dadeng to build an airport has few negative effects on the marine ecology, as that region is not a core ecological zone and is outside conservation areas for the Chinese white dolphin. We conducted a study when adjusting plans in the area, and did so again when adjusting regional plans of sea areas. We conducted further studies when we compiled overall regulations over conservation areas and again on determining the use of sea areas and the environmental impact of the construction of the new airport. After all of these research studies, we finally completed a regional plan for the new airport in Xiamen. The use of sea areas can be very complicated in airport construction, and we drew up plans in advance and redrew them several times on the basis of mathematical and physical models. We also invited several peer researchers to double-check our plans. Some people didn't understand why we needed so many rounds of research, but it turned out that our efforts were in fact necessary.

After going through all the arguments, we also invited the China Consultancy Center on Oceanic Engineering to make a final assessment. We hoped to get this work done before the National Development and Reform Commission approved the project. Environmental impact studies and research on the marine environment are required prior to any construction efforts in any sea area in Xiamen. All influential projects require physical and mathematical models, like the construction in Dongdu port area and Haicang port area. We do not own the expertise required to build all of the physical models ourselves, and therefore we often had to delegate some of the work to Nanjing. Xiamen University and the Third Institute of Oceanography of the State Oceanic Administration were usually responsible for the mathematical models. Basically, small projects usually required only mathematical models, and relatively large ones will require mathematical models through the joint efforts of Xiamen University and the Third Institute of Oceanography of the State Oceanic Administration in accordance with development plan. For years, we have been very cautious about making adjustments to Marine Functional Zoning, because adjustment plans be submitted to the State Council for approval.

City planning and marine planning are on the same level and both require coordination and communication. We always believe that while reviewing coastal reclamation projects, we must strictly control bays, protect natural coastlines and limit the number of sewage outlets. In the implementation of the Marine Functional Zoning, the biggest challenge we met was how to transform existing sewage outlets. It took time to resolve this problem. The original 48 outlets in the western sea areas were not all capable of rain and wastewater diversion, and sewage interception, and so now most have been removed and only 6 remain. These outlets, which are located near old city areas like Navy Pier, Keji High School and Wuyuan Bay, cannot easily be dealt with now, because some of these places do not even have enough space to build a sewage treatment plant.

This is also true for Yundang Lake. There are no major technical problems, but there is also no easy way to solve the problem of sewage discharge, and the key lies in renovating old areas in the city. The reason outlets cannot separate rain water and waste is that every pump station has limited holding capacity and when it rains heavily, pump stations will stop working. So after communication with environmental protection departments and the municipal parks department, we made a plan for the discharge of water in Xiamen Bay. Currently, after we have treated sewage on land on the basis of relevant standards, we then discharge it into the sea. Now integrated management over water pollution in Xiamen Bay has been listed as a national key project in the 13th Five-Year Plan period. Apart from integrated management over 9 rivers, Xiamen has also started to manage pollution in offshore areas.

Dredging sounds easy, but actually it is technologically difficult. It takes quite a long time to determine where to begin. Any mistakes risk destroying the whole ecosystem. Therefore, we must monitor and review the effects after dredging. Sludge cleared out can easily cause secondary pollution and we need to treat it after dredging. As urban construction projects are numerous in Xiamen and require a lot of sand and soil to lay foundations, the cleaned sludge can be put to good use. Now, we have basically finished dredging, except for aquaculture areas which remain there due to historical reasons.

We are now making preparations for a series of projects to build beautiful beaches, islands, wetlands, and coastlines. We improve the landscaping of sand beaches through expanding the distance between beaches, planting trees, and clearing out illegal construction projects. Huoshaoyu Island and Dayu Island are two pilot programs among 17 islands in Xiamen. We will encourage the proper development of marine tourism there. In this regard, Xiaodeng Island is quite ahead of the curve. It combines aquaculture with entertainment, shopping and tourism, and thus improves people's livelihoods. The mangroves in Xiatanwei, Xiang'an District, have been built into the first wetland mangrove theme park in Fujian province. Mangroves there grow well, and can reach the height of an average man. In 2016, Xiamen continued restoring mangrove wetlands, and at the same time, we have also kicked off construction of navigation channels to boost the development of the yacht industry.

We also strive to put a comprehensive safety watchdog mechanism in place. Now, that entertainment activities in sea areas are becoming more and more diverse and some of them are not covered by our management rules. So what shall we do? We conducted some investigations and tried to sort them out in order to develop a proper management framework. Some issues are ambiguous, overlapping with other areas or undefined in laws. The Marine Management Office is in charge of this issue and we are coordinating efforts. Now the largest problem we are faced with is industrial supervision and the second is local supervision. For example, for a sailboat under 5 meters in length, the Sports Bureau, Marine Bureau and Port Bureau each have their own regulations. So when we are planning for coastal tourism programs, we must cooperate on how to ensure proper security and management. Industrial supervision efforts were meant to fall to each district, but the districts lack the proper institutions. So what shall we do? We coordinate. Since its establishment, the Marine Management Office has played a very positive role in this regard, probably the most positive nationwide.

In the past two decades, rapid economic development in Xiamen has greatly promoted marine protection, because without economic development, marine environmental protection could not have been possible. Without money, you can do nothing and vice versa—if you do a good job in coastal management, you not only protect the marine environment, but also boost the economy. Some always argue that one or the other should be our first priority. I actually believe that economic growth and marine protection are mutually complementary and enjoy equal priority, and we can work on them at the same time. It turns out that Xiamen has not only achieved rapid economic growth, but also has managed to protect its marine environment, or even improve it. Xiamen's experience in returning farmland to seas has drawn national attention and has been promoted nationwide. Xiamen's experience in Integrated Coastal Management based on the principles of marine ecology and international marine exchange and cooperation also serves as a role model for the rest of the country.

※ Management of the Jiulong River is also very critical

When discussing Integrated Coastal Management in Xiamen, we cannot skip over management of the Jiulong River, which is the main task in the third stage of management.

At first, we did not have a complete understanding of the level of pollution in the Jiulong River. We began to notice that every time there was a flood, the seas were filled with waste. Earlier, dinghies were used for cleaning. After Vice Mayor Pan Shijian took charge of the Xiamen Municipal Bureau of Parks and the Municipal Bureau of Oceans and Fisheries, he approved the construction of two large ships for cleaning waste on sea. Then some experts proposed that the Jiangdong Dam Bridge and Xixi Dam Bridge in the upper streams could possibly be used to intercept waste. Jiangdong Dam Bridge was under our jurisdiction, but not Xixi Dam Bridge. So we hoped to start with the Jiangdong Dam Bridge. Some suggested setting up nets at the estuary of the Jiulong River to protect Xiamen, but it was not plausible. So we wanted to join hands with Longhai, a city under the administration of Zhangzhou, Fujian Province. With support from the government, funds were not a problem. But due to some problems in Longhai, this effort was suspended, but remained subject of discussion among departments.

Our action plan for management of the Jiulong River had evolved into a project of regional cooperation and eventually led to an important decision of the provincial party committee. We traced marine pollution back to its source—pollution from river basins. So we needed to first make an action plan and implement it in Xiamen. During implementation, Xiamen established a partnership with Zhangzhou and Longyan through economic coordination. So our action plan was elevated from the municipal level to a regional intervention and then to the provincial level and eventually led to a resolution. The joint efforts of Xiamen, Zhangzhou and Longyan have led to many outcomes since then.

※ Legal education should start with children

Sometimes, a minor change in the seas should set off alarm bells.

For example, when short-necked clams appear on beaches of the Huandao Road, it means there is siltation which has changed the habitat of lancelets. This would alarm us to monitor the waters to estimate the total number, size and density of lancelets in the surrounding area. If the environment is no longer suitable and the lancelet population has decreased drastically, we will adjust our protection measures. Every year, we publish a report on marine environmental conditions which assesses protected species in conservation areas. We also strive to stop people from collecting short-necked clams on beaches based on Administrative Measures on Natural Conservation Areas of Lancelet. To solve this problem, first, we organize a team of volunteers and divide them into groups of six to talk to those who collect short-necked clams and persuade them out of it. Another measure is to put up a sign board to warn against it. We have also begun efforts at scientific research on artificial breeding of the lancelet, and once we succeed in increasing its reproduction, we will release lancelets into the wild.

In protecting the Chinese white dolphin, though we had adopted *Regulations on the Protection of Chinese White Dolphin*, we met with some problems in practice. Construction areas for some projects were located just outside of the nature reserves and we had to take some precautionary measures when construction was underway. But what measures and how could such measures best protect the Chinese white dolphin? We once studied the sounds dolphins make in order to find ways to protect them. For example, we arranged several ships outside of wherever there was expected to be an explosion, and when the explosion was about to take place, we would pound on the ships to make sounds and drive dolphins away. Due to these measures, underwater explosions caused no major accidents to the Chinese White Dolphin.

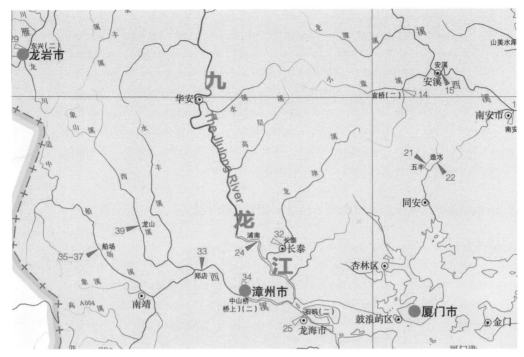

▲ A map of the Jiulong River Basin

These are important elements in building a sound legal system. In addition, we also need regulatory documents to instruct people what to do. For example, all construction firms needed to learn the proper procedure for chasing away dolphins prior to underwater explosions. The Xiamen Municipal Bureau of Oceans and Fisheries also sent officials to instruct and help with this. So in "law" enforcement, enterprises had rules to comply with and standard procedures to follow. This was real progress.

Gaining public support for laws and regulations has been a focus of our work since the establishment of the Marine Management Office. The Coastal and Ocean Management Institute of Xiamen University evolved out of the then training center of the Office. We once considered where to conduct training: Xiamen University or perhaps the Third Institute of Oceanography of the State Oceanic Administration or Fujian Institute of Oceanography, or other colleges or universities. Considering the professional requirements and the need for opening-up, we decided to put training programs under PEMSEA. Consequently, the municipal government and Xiamen University jointly established the International Training

Center for Coastal Sustainable Development. In our first training program, we invited Mr. Qu Luoping, the first Director General of the National Environmental Protection Bureau to lecture for governmental officials. This program was such a success that it helped greatly to promote the training center.

We began using the media to help report on our efforts from quite an early stage, for example, through TV news programs like Focus, Viewpoint on Oceans, and weather forecast. For us, the key Xiamen TV program is still Viewpoint on Oceans. At the same time, we also use print media for publicity. For example, we have a fixed column in the Xiamen Daily, and we continue to publish articles regularly. Additionally, we are also making attempts to harness the new media by recruiting professionals with experience in the news industry so as to enhance public understanding and support for our laws.

In the early years, we even directly took to the streets ourselves to do publicity, by setting up street stands. As soon as new regulations were adopted, we would go to crowded places like ports to publicize them. When the Huandao Road had just been completed, we erected signs and billboards on the beaches. Though they were easily worn out and had to be replaced very often, they played a positive role in showing our efforts to the public. We also made trips to small villages to publicize our laws and regulations. For example, we went to villages that exploited sands to publicize regulations on sand exploitation. We went to villages engaged in aquaculture to publicize the new bans and restrictions on fishing and aquaculture. We have different methods for increasing publicity, like creating catch-phrases, sending out booklets and leaflets, and presenting small gifts to households. However, sometimes when people's lives are affected, it can be difficult to get the public on our side nevertheless. Some reckless people resort to force and obstruction, for example by putting cars in the way of our construction or campaign efforts. This highlights another flaw in our legal system: lack of regulations on punishment for illegal acts. But if you view it from another perspective, you might find this issue to be outside of the scope of our responsibilities because it is about public security, not coastal management. So we are still exploring ways to raise publicity and public support for our laws and regulations.

Apart from publicizing and issuing guidance on our laws and regulations, education on ecosystems of oceans is another focal point of our publicity efforts. In recent years, we have joined hands with the Xiamen Municipal Bureau of Oceans and Fisheries to build education bases for raising public consciousness about oceans and marine ecosystems. Although, compared to the rest of the nation, Xiamen has many more such bases, we still don't think that it is adequate, and will invest more at the municipal level. We also use scientific research ships and the Underwater World as our bases for publicity. For example, at Underwater World, after children watch the shows of dolphins, we hand them booklets on how to protect dolphin species, their ecosystem and their food chain. We also established a partnership with Tianxindao Primary School in Haicang District and provided the school with materials on oceanography so that the school can compile it into textbooks and set a course on oceanography.

With the development of new media, new methods are now available to educate the public about the marine environment, management efforts and laws. In addition to cartoons, Xiamen Municipal Bureau of Oceans and Fisheries also spread information through WeChat, for example, by organizing a marine quiz show, where citizens with top scores were invited to visit the Chinese White Dolphin Rescue Base. The Municipal Bureau of Oceans and Fisheries is said to consider turning this kind of activity a regular program by introducing a credit system.

In recent years, Xiamen Municipal Bureau of Oceans and Fisheries' efforts to promulgate and publicize laws have been widely recognized by the masses and superior departments. In June 2016, Xiamen Municipal Bureau of Oceans and Fisheries was presented the "Award for Outstanding Group in Raising Public Legal Awareness" during the sixth five-year law publicity period from 2011 to 2015 by the Publicity Department of the CPC Central Committee, the Ministry of Justice and the National Law Publicity Office, the only marine department to win such an award.

※ Playing a leading role in the international arena

Since 1994, the Chinese government has joined hands with international organizations like the UNDP in setting up pilot programs in Integrated Coastal Management in Xiamen, and we have begun joint projects with with PEMSEA. From 1994 to 1998, Xiamen began the first round of Integrated Coastal Management. In 2001, we started the second round. From pollution control and management, to ecosystem-based marine environmental management, to basin management, to the PPP projects, we have kept moving forward. Now we are promoting the development of the marine economy. We can see the integrated coastal management efforts in Xiamen are bearing fruit. What will the future bring? We don't know. We will just keep exploring new avenues for practical problems.

The Integrated Coastal Management in Xiamen plays a leading and exemplary role on the international arena. Before 2001, PEMSEA founded the Regional Network of Local Governments (RNLG). After 2002, the PEMSEA Network of Local Governments (PNLG), which serves as a forum for exchanging information and experiences in integrated coastal management practices among local governments in the East Asian Seas, started to take

▲ Xiamen has always been known around the world for its unique fishing resource of amphioxic. We are educated since childhood to protect amphioxus, and we have been taking action all the time.(Photo / Nian Yue)

shape. Now PNLG has 42 members and its Secretariat is set in Xiamen.

As Xiamen boasts experience in Integrated Coastal Management, the office of the APEC Training Center on Sustainable Development is located at the Third Institute of Oceanography of the State Oceanic Administration. Training workshops for ministers from third-world countries and commissioners from developing countries also choose Xiamen as their venue. This work is conducted by the Fujian Institute of Oceanography. Some people ask why the Marine Management Office doesn't participate in these efforts. What we believe is that if others are capable of doing it, we should give full play to their capabilities, and we don't need to do everything ourselves. We look at the whole picture and organize the right institutions to do the appropriate jobs. This concept also applies to our marine projects.

The fundamental difference between the Southern Marine Research Center and the Northern Marine Research Center is that we, the Southern Marine Research Center, give full play to the collective advantages of local research institutes and schools, like Xiamen University, the Third Institute of Oceanography of the State Oceanic Administration and Jimei University. We pool their resources together and the Office will organize and delegate duties for each. For example, Third Institute of Oceanography is mainly responsible for marine biopharmaceutical medicine and its applied research, a kind of industrialized applied research. Xiamen University focuses on fundamental scientific research and the Third Institute of Oceanography of the State Oceanic Administration conducts further research based on their findings. Eventually, outcomes are shared, which is beneficial for industry, research and the sharing of resources. As there are no conflicts of interest, we can easily gather people together and work together.

The Marine Management Office applies this model to its day to day work. For example, in the past, aquaculture was not under the administration of the Office. We experienced some difficulties in its management when it was covered, but later on after phasing out aquaculture, this problem was resolved and it became easy to coordinate.

Though we no longer have a substantial fishing industry in Xiamen, it doesn't mean we have no fish to eat. We encourage people to build ships and catch fish in other areas and sell in Xiamen, or raise fish outside and supply the market here. These ways have both pros and cons. How can we encourage people to do this? The key is about making correct policies and creating an enabling environment. In the past, we granted large subsidies for building ships. But after two years, these ships produced no substantial fish catch. So we adjusted policies and added the condition that we would only give subsidies after ships came back with fish. The policy worked and while ship building slowed in Xiamen, fish became more and more plentiful in local markets. We plan to make further adjustments to our policies in an effort to build Xiamen into a hub of seafood products nationwide and worldwide. As Xiamen features convenient transportation by land, sea and air, we also hope to develop cold-chain logistics for seafood products in Xiamen.

In Integrated Coastal Management, one important way to encourage public participation is through training programs. The State Oceanic Administration, the municipal government of Xiamen and Xiamen University jointly founded the International Training Center for Coastal Sustainable Development, predecessor of the current School of Marine Affairs of Xiamen University. In the past, the Center provided regular training programs for officials

▲ In June 2006, the Global Environment Fund sent a reporter to Xiamen to report Xiamen Coastal Zone Integrated Management.

from ICM parallel sites and demonstration sites in the Eastern Asian Seas Region and for heads of departments of the various municipalities. Later on, the school began to accept undergraduate students. The Fujian Institute of Oceanography is also working on developing training programs, mainly targeted at third world countries.

Xiamen's experience with ecological rehabilitation has been promoted nationwide. Integrated management for ecological rehabilitation has developed a solid methodology, consisting of the establishment of management mechanisms, scientific support, integrated environmental management, publicity efforts, building of legal frameworks, construction of the ecological economy and "ecological civilization". We have established a robust mechanism for protecting threatened species such as the Chinese white dolphin, lancelet, horseshoe crab and Chinese egret. We are pioneers in promoting the development of the marine economy and also national leaders in the production of marine pharmaceutical medicine, of which we are very proud.

Since around 1990, the Integrated Coastal Management mechanism in Xiamen has matured and gained international recognition. The results of our practices and mechanisms are coming to fruition. The Yundang Lake, which once tarnished the image of Xiamen, has become one of the most successful cases of coastal management in Xiamen. Xiamen actively promotes its management mechanism and so we launched the Xinglin Bay project and Wuyuan Bay project later on. The Xiamen municipal government has also been presented an Award for Excellence in Coastal Governance by PEMSEA, jointly founded by the GEF, UNDP and IMO.

▼ Since its establishment in 2005, Xiamen International Ocean Week has developed into a marine cultural festival widely attended by the public, and a platform for exchange on global ocean policies, science & technology, decision-making and action. (Photo / Wang Huoyan)

One of the biggest differences between our program and other international projects is that we are exploring new solutions to marine pollution. In the second stage, as we proceed with Integrated Coastal Management, we turn our focus on basin management and the prevention and control of waste on the sea.

Integrated Coastal Management can be applied to China's marine cooperation with ASEAN. Now we have expanded our international partnership, from within the Eastern Asian Seas Region to within ASEAN, with plans to even expand to Europe. The PNLG of PEMSEA, which has become an important international organization, helped to spearhead some of our international cooperation efforts. After the founding of China-ASEAN Marine Cooperation Center, it will serve as a new platform for international cooperation. In the next step, we need to think of how to best develop this platform and how to create new opportunities with it.

We have experienced several stages of development in the past 20 years of integrated coastal management in Xiamen. The key factor for our continued progress has been innovation in systems and mechanisms. Additionally, creating a strong supporting legal framework has also proved very important. The significance of systems and mechanism is that they enable the building of a strong and authoritative management system, with which we are able to resolve problems.

When I graduated from university, Xiamen was not my first choice of work place. After decades, although it fails to meet the ideals of my imagination, it remains a wonderful city. When I walk down the street and breathe in the salty sea air and enjoy the gentle breeze, I feel like I live in the city and at the sea at the same time. Xiamen, with its clear seas, blue sky, sand beaches, yachts, sailboats, dolphins and sea gulls, is very beautiful if you take an aerial picture of it. Many say that people in Xiamen lead a very comfortable life and, as a result, they don't have the motivation to work hard. But I don't agree. I believe people in Xiamen are still motivated and hardworking, as they have sea water in their blood and retain some habits of fishermen. They go fishing every day and no matter how much they catch today, they will go out again tomorrow. I can tell many young people in Xiamen are courageous, ambitious and hardworking. If we can provide them with an enabling business environment, they will have a sea of opportunities to start up their own businesses. Though we must keep in mind rules and regulations, with less red tape, the world will be their oyster.

Now, when I swim in the sea and listen to the clapping of waves on the ocean, it feels like I have returned to the sea I was familiar with as a child. Seeing fish swimming by and sea gulls hovering overhead, I feel very comfortable and at ease.

Collaboration Is the Core of Integrated Coastal Management

Ruan Wuqi

Ruan Wuqi could have had an entirely different life but for the Second Taiwan Strait Crisis.

In 1958, the crisis broke out. All fishing boats in Xiamen, the frontline of encounter, had to be moved to the South China Sea. "I was just beginning middle school, so I stayed with relatives in Xiamen instead of moving to the South China Sea along with my family. " One year later, his family returned to Xiamen. Ruan had intended to become a fisherman after finishing junior high school, but as he managed to test into a high school, his father allowed him to continue his education. As a result, Xiamen Harbor loses a humble fisherman but gains an honorary director of the Fujian Institute of Oceanography.

Although he did not become a fisherman, Ruan still held a deep affection for the sea as a child born to a fishing family in Xiamen Harbor. Due to his love of the sea, he chose to major in marine chemistry at Xiamen University, and due to his attachment to Xiamen, he took a job in a marine expert panel when the city's Integrated Coastal Management (ICM) was in its infancy. He has worked hard with his colleagues over the last two decades on the"Xiamen experience" in Integrated Coastal Management—an achievement made possible by several generations of Xiamen's marine experts and government officials. The experience may not have been directly adopted by other coastal cities, but still assumes great significance. Talking about the city's ICM today, Ruan exclaimed, "This is no doubt the greatest example of ICM in China and even the whole world!"

"When the moon falls and the tide ebbs, Yundang Lake is dotted with the lights of fishing boats. "—these two lines depict"lights of fishing boats on Yundang Lake", one of the eight"best scenes in Xiamen" in the past. As Ruan Wuqi recalled, in his childhood, he would take shelter from the wind when fishing with his parents. When night fell, Yundang Lake became a haven for small fishing boats, whose lights, and melodious flute players, formed a picture of harmony between humans and nature.

Recalling Xiamen's past, Ruan Wuqi noted, "This is our memory of the bygone days in Xiamen's sea. It is unlikely that we can restore the scene of 'lights of fishing boats on Yundang Lake'. But when building Xiamen into a beautiful city, what we can do is to use ICM to turn Yundang Lake into a gorgeous postcard for Xiamen, a city surrounded by the sea, and make Xiamen's coastal zone a symbol of the city's prosperity!"

※ A son of the sea working for the sea for nearly half a century

Born to a fisherman's family, I spent my childhood living on a fishing boat with my parents, because we had no money to buy a house, and my parents, who fished on the sea throughout the year, would rent a house when going back on land for a short time. After birth, children in my family would stay on a boat along with our parents and go fishing. As I remember, when our parents were busy fishing and had no time to attend to us kids, they would tie us to the boat with a rope lest we fell into the sea. When we got tired, we would sleep in a bamboo basket covered with a tattered quilt, just like puppies.

At that time, fishing was a humble profession, and fishing families were impoverished. I had no shoes to wear until junior high school. But I still took pleasure in my poor childhood thanks to the company of the sea. I was interested in everything related to the sea. My grandfather was the first fisherman in Xiamen Harbor to buy a barometer to help predict typhoons. He did not know how to read the numbers but could forecast typhoons based on how far the pointer moved to the left or right—which to me seemed like magic. When I grew up and learned about the relevant science, I knew it was nothing but a barometer. Wind, rain, waves, fish, sea gulls ... they accompanied me throughout my childhood and gave me endless fantasies.

Out of my love for the sea, I chose to major in marine chemistry. Those who spend most of their time on the sea know that the purpose of sending their children to university is to keep them away from the sea, or "to keep the hands and feet dry and away from the seawater" in their words. But when I graduated from university, I returned to the sea to continue my story.

After graduation, I was assigned to work at the Wenzhou Branch of the Marine Fisheries Research Institute of Zhejiang. Ten years later, in 1974, I was transferred to the Subtropical Plant Research Institute of Fujian in Xiamen. Though unable to work in marine research, I was still content to work in my hometown.

After 5 years at the Plant Research Institute, Lu Weite, the then director of Fujian Science Committee, proposed establishing a marine research institute and started to bring together marine professionals. As a result, I began working at the Fujian Institute of Oceanography at the end of 1979—returning to the same industry as my major, or "rejoining my unit" as it was called.

A prevailing slogan at the time went like this, "Study hard, stay healthy, and work for half a century for the CPC!". Having worked for exactly 50 years from the very beginning to the end of 2016, I managed to fulfill my commitment. Out of the 50 years of my career, 45 were devoted to the sea. Because of this, and due to my fisherman's roots, so I was called by colleagues "son of the sea".

※ Concerted efforts to protect Xiamen's lifeline

Today the concept of marine protection has been engrained into the hearts of Xiamen people, but 20 years ago, few realized that their living habits could harm the sea. Why should we protect the sea in Xiamen? Because Xiamen is a coastal city, and the sea is its lifeline. Those living in this city would not want the lifeline to break. The concept was first realized by marine professionals, who, together, appealed for *Provisions on Marine Environmental Protection of Xiamen*. But at first, their efforts did not go well, due to the lack of specific plans, laws and regulations to restrict harmful behaviors; plus, the

Introduction to the speaker

Ruan Wuqi, former director of the Fujian Institute of Oceanography, former head of Xiamen Marine Expert Panel, has participated in the Integrated Coastal Management in Xiamen for more than 30 years.

government and the public were used to this unsustainable way of living.

Coastal management used to be conducted by various departments and industries. When it came to development, agencies mostly focused on expanding their own revenue rather than collaboration, while when it came to management, they only thought about expanding their own power. For example, expanding aquaculture did boost economic growth in China, especially in rural areas. As the production methods were backward back then, producers tended to resort to excess pesticide usage when diseases broke out, damaging the entire marine ecosystem. Today Xiamen Island looks like a piece of moon cake, but in the past Xiamen had a harbor enclosed by another, and a zigzag coastline, which was indeed beautiful. Obtaining resources from the sea is not a mistake itself, but doing so in a haphazard way can lead to endless troubles, especially to the best stretches of the coastline where exploitation and usage involve low costs.

Our notion of protecting the sea used to be superficial and partial. We would clean up particularly polluted areas, but could not thoroughly solve the problems, since seawater flows continuously, and Zhangzhou, Quanzhou and Longyan, upstream of Xiamen on the Jiulong River, all discharged sewage into the rivers that flow into Xiamen. Seeing garbage piled up along the river banks during our inspection of the Jiulong River, we asked a farmer, "Why do you leave the garbage unattended by the river?" "It's ok." he replied.

1. On May 31, 2003, the Xiamen marine functional zoning revision meeting was held.
2. Ocean & Fisheries Bureau of Xiamen staff studied Xiamen marine functional zoning map.

"When it rains, the trash will be washed downstream. " It is fair to say that our prior coastal management efforts were not management in the real sense, but nothing more than the blind application of land management policies to the sea, achieving no results due to the lack of objectives and coordination among the various marine departments.

In 1997, Xiamen officially began its ICM.

The UNDP has an ICM project, later known as PEMSEA (Partnerships in Environmental Management for the Seas of East Asia). The UNDP official Dr. Chua Thia-Eng, whose ancestral home was Tong'an, was deeply concerned about the marine environment of his hometown. Considering the abundance of marine experts gathering in Xiamen, he believed this city was capable of conducting a PEMSEA project. Compared with the mere 900,000 US dollars given by the UNDP, we had invested much more money in the city's coastal management, but had little to show for it due to a lack of synergy. When he asked for my opinion, I said straightforwardly that the fund could only be used to initiate projects, but that the success of the projects relied on the creation of synergy between all departments in Xiamen in the name of the UN. Amazed by my answer, he said I was the only one who had given him an honest answer among all directors of the research institutes.

The ICM pilot program began in Xiamen in 1997, with mayor Zhu Yayan directly leading the Marine Expert Panel. That year, Coastal Functional Zoning (CFZ) was implemented as a principle for all industries to comply with in the utilization of marine resources. Afterwards, when we attended a meeting in Thailand, Xiamen's ICM was highly commended.

※ Collaboration is the core idea of ICM

The urban development of Xiamen is the epitome of China's marine economic development. For example, some suggested reclaiming land from sand beaches, valued at RMB 400 million, at Xiamen University. That is not a large sum of money nowadays, but back then it was quite a lot. After some debates, the expert panel members decided to stick to the principle of sustainable use of marine resources. Commercial developments on the reclaimed land could indeed make lots of money, but if the sea were not turned into land, it could be used for aquaculture, yachting, building ports, tourism and so on. The sea is special because it is cross-functional, but marine resources disappear forever upon land reclamation.

When doing CFZ, we first needed to determine the main function of a sea area. A group of experts carried out this task under the leadership of a exploratory team headed by the mayor. The first step was to break the rules set by the State Oceanic Administration (SOA) and focus on the real situation in Xiamen. Some commented that the CFZ was successful because it was conducted on the basis of the realities of Xiamen, not in an officious, mechanical way. After several years of adjusting the CFZ, its effects became obvious.

Collaboration is the core of ICM. When I gave lectures to foreigners, I often talked about this. Why? Collaboration was difficult, because the former management mechanism was divided among different departments and industries. Few people knew that in 1983, an ICM pilot program failed in Liaoning, as the relevant institutions failed to coordinate their efforts. Unprofitable tasks, of which nobody wanted to take charge, were left unfinished, while agencies fought to snatch up profitable opportunities, causing great overlap and conflict. In this regard, the Xiamen government broke the rules and set up a collaboration mechanism of the Marine Management Office, so as to fully exercise the authority granted to it by the local government. This mechanism was able to create collaboration step by step.

Throughout the implementation of ICM, the municipal government and the municipal CPC committees played a key role in fostering a cooperative atmosphere. We often talked about the important role leaders play in practice. Over the past 20 years, we have witnessed the charisma of the municipal leadership in the city's ICM. The ICM was a high-level administrative task. What was the administrative mechanism like back then? The working environment depends on the leaders. I once told Zhu Yayan in person, "Mayor Zhu, every time you did a field research, the directors would follow you; but if a director paid a visit, the relevant institutions just send an employee to accompany the inspection. For subordinates to properly carry out orders, we needed to establish a hierarchical administrative collaboration mechanism. " Clearly aware of this situation, Mayor Zhu emphasized the importance of interdepartmental collaboration at every meeting. At an expert panel meeting held by the municipal government, I was upset at complaints made by the directors of relevant departments about our team. I wrote a note to Mayor Zhu, asking for 10 minutes to explain our ideas. He responded with a note, allowing me to sit beside him at the meeting and speak as long as I wanted. At the meeting, I explained the entire situation thoroughly to those officials, otherwise they would continue to think we experts were just making trouble. Actually our work was for the benefit of the orderly development of the sea in Xiamen. Officials should not focus on the aquaculture business or the vegetable basket project just for their own political gain. What would the transportation department do if the sea channels were blocked? So the key has always been collaboration.

Leaders and experts talked about whatever they wanted to at every meeting. Mayor Zhu often asked me, "What kind of absurd theories are you going to talk about today, the wild child from Xiamen Harbor?" The leaders made such jokes because they were already close to us experts.

Busy at work on weekdays, Mayor Zhu would sometimes choose to review our work on weekends. He frequently told me, "I will treat you to dinner on Saturday evening. " I replied jokingly, "That means we will have work to do. " There were no secrets between us. If we met with any problems, we could directly ask him for help.

After the completion of management mechanisms for Xiamen's leadership among various industries, government agencies, qualification institutions, and the Marine Expert Panel, coastal management began to speed up. As members of the Marine Expert Panel, we were clear about our responsibilities—we are those who should provide "ammunition", conduct field investigations, and provide the first-hand materials for the leadership's reference in decision-making. Such a division of labor laid a solid foundation for the entire integrated management. According to a UNDP official, the seawater of Xiamen, quite dirty five years ago, has become much cleaner during his most recent visit, which means the integrated management has been effective. This is attributed to the government's leadership and the public involvement.

※ Understanding the essence of the "integrated management"

Coastal management should be "integrated" rather than separated. What does this mean exactly? It concerns the integration among government departments, and between science, technology and management. I put forward a viewpoint that in the ICM, we should take the management department on board when doing all scientific research or technological work. Any problem the department encountered should be handed over to professionals and technicians for solution. Without cooperation between management and research institutes, the project would not go well. That was why a host of research projects conducted later in Xiamen, such as for ecological restoration, have all been reported to the management department. Just imagine, in the process of marine governance, could you win public trust if you could not provide a scientific basis for management decisions? Even the government departments would disapprove, and might believe marine authority to be in chaos. With heavy responsibilities on our shoulders, our Expert Panel have to provide reasoning for any

1. On December 19, 2001, representatives of Xiamen Municipal People's Congress visited the West Seas before rehabilitation.
2. On June 22, 2002, City Leader inspected the west seas, standing on a fishing raft.
3. On October 11, 2002, Xiamen CPPCC members inspected the rehabilitation work of the west seas.

action. For example, in the case of banning aquaculture, we need to explain the legal bases, the actual status of the effort, the compensation for fishermen, job training programs and profit estimates. If the management departments have trouble answering those questions, they could leave it to our team. Then we would assign the job to different departments in Xiamen, and review the collected work and solicited opinions. We keep improving our answers until they are satisfactory, and then submit them.

Such collaboration exists between managers and technicians. During that time, our team frequented the seas to check the status quo of our initiatives. This has been a useful method during our ICM.

Against this backdrop, in 2003, a marine law-enforcement team was established in Xiamen—the first such team in China that was made up of public servants. The team unified law-enforcement efforts and handled issues as they arose, rather than wasting taxpayers' money by idling on the boat every day. I used to attend marine work meetings on the issue of enforcement in several provinces and introduced our team's work model. Only Xiamen has fully realized an enforcement mechanism. We are very much ahead of the curve in our work, and our results served as an example for other cities. For this reason, we have won many awards for our marine work.

One of most startling events in Xiamen's coastal management was the demolition of the Sea View Building near Xiamen University in Siming District. The building had been constructed on the beach, leading to continuous erosion. It was a profitable tourist attractions in the district but from the perspective of coastal protection, it had to be removed. After two or three years of appeals, we finally went through with the demolition, and the lost sand returned.

There were plenty of cases similar to the Sea View building case. We fixed problems wherever they cropped up, part of the first step of ICM. For the second step, the local government proposed building Xiamen into a coastal city, and later a "marine ecological civilization". Our ICM kept pace with these goals, allowing other tasks to be carried out, including ecological restoration, fry releasing, the planting of mangrove forests, ecological rehabilitation, etc. Only when the departments involved in coastal management had their own policies could the sea be effectively managed, its use no longer free of charge, disorderly or unlimited. As the city's marine ecological restoration continued, we came up with an ecosystem-oriented management concept. When we put forward this concept at the World Ocean Week in Xiamen, we were met with resounding applause.

Coastal management in Xiamen not only keeps pace with the international trend, but also meets the city's needs for economic and social development. The advanced measures we propose guarantee the completion of the strategies set by the city's authorities.

※ Transitioning from partial to integrated management

We have upgraded our reactive, fragmented marine environmental protection into a proactive, integrated management scheme for the entire marine ecosystem. Whether these measures can work in other cities is a question, but there is no doubt our efforts are useful as reference for other cities hoping to conduct this work.

Throughout Xiamen's ICM, scientific support and interdepartmental collaboration have been aimed to promote the city's sustainable socio-economic development. Without this concept, the government would not recognize our work. Xiamen establishes the Southern Oceanic Research Center and Southeast International Shipping Center for these reasons, and also to promote international cooperation, and now they also serve as a response to the "Belt and Road" Initiative.

For example, the 5 berths at Dongdu Port have been all transformed into the home port for ferries in order to promote the ferry industry. The port has not yet turned a profit, but has great potential to bring economic benefits. The transformation also improves the environment of the whole sea, creating a better environment for white dolphins in the core reservation areas with fewer ships passing by. Such changes help adjust Xiamen's economic structure and create a better marine environment.

Our Expert Panel has prepared review plans for each project. For example, we once reviewed what impact did the Dongdu Port Ferry Project have on white dolphins, how and to what extent and what measures should be taken to alleviate the effects. We would then use data to analyze the possibility of white dolphins showing up in the sea area over the

▼ At midnight of May 6, 2003, law enforcers of fishery, fishery supervision and maritime supervision departments of Xiamen intercepted an attempt to tow a fishing raft to Tong'an Bay at the mouth of Gaoji Causeway entrance.

next decade. There were two distinct opinions: some said white dolphins used to exist here but no longer come to the area, so there is no need to protect them; but I was of the opinion that they disappeared because of the damaged environment, so we must protect them. Often faced with a divergence of views, we had to make compromise. Over the past 20 years, we marine experts have held a quite firm position. We work solely to promote sustainable social and economic development in Xiamen. We need to keep pace with development and even look to the future, addressing problems facing the environment and ecosystem in the city, to ensure the vitality of the ICM program.

In addition, throughout the ICM, the municipal government has been seeking public participation. Viewpoint on Oceans is a program on XMTV that enjoys good viewership. I have been on that program several times, and often played its episodes to educate my family about the sea. The public response to the opinions we express on TV also serves as a gauge of public support. This is also a way to improve our efforts. We must not be afraid of answering questions from the public. If we sometimes could not understand the questions raised by the audience, we would frankly admit, "This is not my research area, but your question will prompt me to learn more about it". Audiences appreciate such candid responses. Interaction is a key channel for public participation.

At the same time, the government has used questionnaires, road shows, proposal signing ceremonies and other measures to demonstrate the coastal management efforts to the public. When building Yanwu Bridge on the Huandao Road, after the municipal government listened to public opinion, it was decided that the bridge should not obscure the sight of Gulangyu Island, a decision that won praise from the public.

We also put much effort into educating our kids. Take myself as an example. My grandchild's kindergarten once invited me to talk about the sea to the kids. Since oceanology is a set of profound theories, I was clueless about how to prepare. That day, I braced myself to deliver a 40-min speech, sweating all over. But that evening, I received a phone call from a close government official, whose son was in the same class with my grandchild. He said that his son told him about an old man who advised people not to litter in the sea to ensure it would not get sick, and he guessed that the old man was me. I was delighted to see the response. I believe, in our ICM, the publicity work was done rather well. Actually public participation does not have to be viewed as a show or a burden, but is actually a means to enhance people's awareness of environment protection. This is the idea we have been upholding over the last 20 years.

※ Technology-enabled ICM

How technology has supported our ICM is also worth noting.

Technical support in Xiamen outperforms that of other cities. When some laws and regulations introduced in Xiamen encountered problems during enforcement, we conducted special research in response, and I participated in the initiation of a dozen related projects. At the early stage, we did research only to provide scientific bases for programs; later, we did research to facilitate the enforcement of laws and regulations. This was very unlike doing research at an institute and completing it for the purpose of publishing papers. Throughout ICM, the research results would not be published, but they did help in the enforcement of marine laws and regulations introduced by the municipals government and CPC committee. Research subject must be selected properly, otherwise it would be useless. And we required all researchers to work on providing bases for law enforcement instead of doing research for research's sake. When I took part in the Tong'an Bay project, we were asked to define the areas suitable for aquaculture. As for the the function of the western sea area, we had been arguing for a decade about its primary function—transportation or aquaculture. We did not completely resolve the problem until we began work on another project on how to determine the main function of a sea area.

At that time, Tong'an Bay was mainly used for aquaculture in accordance with Marine Functional Zoning. We started a project and divided it into 3 parts for different institutions to accomplish separately—fisheries for the Marine Fisheries Research Institute, hydrodynamics for the Third Institute, and marine environment for Xiamen University. Our Expert Panel only acted as a bridge between the marine research institutes and universities in Xiamen, and was responsible for reviewing their work. We led the project, and distributed our funds to those three institutions based on their needs. I told them I would reserve 20,000 RMB for the review process, and the remaining funds could be distributed to them according to the rules. The project went quite well.

While doing our job, we noticed changes taking place in the entire marine environment in Xiamen. We did not have any specific research projects to study these changes, as we hoped to sum them up from our long-term research. This sum-up task was assigned to me after my retirement according to Xiamen's Marine Environment Protection Plan issued by the Fujian Association of Senior Scientists and Technicians. I was still sailing on the sea when I received this task. After the project was finished, I kept urging the SOA to finish the review. Because new policies and regulations were issued on a daily or weekly basis, I would have to track and revise them every day. At last, during the review, those review experts told me they would not give any disagreement and understood that how hard I had worked, and that I even rewrote some parts of the project when someone else had failed to write them correctly. I firmly believed that our city's marine environmental protection must achieve a result better than other cities (or at least one satisfying to me).

A rigorous approach to science has permeated the whole ICM process. When implementing Marine

▶ Gulangyu has been transformed from an ordinary residential area into a scenic spot. While creating more income for the city, Gulangyu also demonstrates the positive results of comprehensive rehabilitation of Xiamen coastal zone management.
(Photo / Wang Huoyan)

Functional Zoning, if we were not certain about some issue at the management level, we would go on a field trip to determine the actual situation on site. I believed every marine researcher to be strong in their own field and willing to study the other sub-disciplines of oceanography, considering its complexity.

Last but not least, our system of communication and cooperation is also quite unique. Apart from the World Sea Week, our view has extended beyond Xiamen's 390-square-kilometer sea area to the broader Xiamen Harbor in other ways. Our marine enforcement mechanism is now under the ambit of the Xiamen-Zhangzhou-Quanzhou enforcement alliance. The most difficult part of ICM is cross-administrative-district management. By holding joint meetings for marine law enforcement in Xiamen, Zhangzhou and Quanzhou, we hope all people can realize the importance of environmental protection no matter what the results. This move is rarely seen throughout the country.

I have been asked by some people if I have any regrets about my time working for ICM. Of course I have a few. Some of our tasks were not completed despite good starts; the current marine enforcement alliance should strengthen ties with neighboring cities along the the Jiulong River. Our goal is to protect the entire Xiamen Harbor. The upstream part of the Jiulong River accounts for 60 to 70 percent of the pollution coming in to Xiamen Harbor. I once suggested dividing this river into several stretches—from Longyan to Zhangping, and from Zhangping to Hua'an—and setting up two collaborative teams responsible for each stretch. But two questions have thus emerged: who should be the organizer and who should provide fund? This task has yet to be completed.

Xiamen's marine environmental protection is far from complete, as the water quality in a number of sea areas still fails to meet standards. For future efforts, in order to proceed further with coastal management, we should focus on some advanced topics. Certainly we have done pretty well over the last few years, but should do more to help the government propose suitable items when drafting the relevant laws. When we were cleaning up the western sea areas, where all the fishing rafts owned by my family relatives were located, I mobilized the public to give up aquaculture on TV. When I returned home, I was scolded by my relatives for valuing the government over my family's interests. I told them they had it wrong. What kind of future would we have if we lived on the rafts generation after generation? Just think about the kids. When I was young, even my illiterate parents sent me to school to change my fate. For the sake of our offspring, we must protect the sea and bequeath to them the gifts of nature.

Coordination Between Agencies Is Essential for Coastal Management

Lin Hanzong

▼

Born on Gulangyu Island and raised in Xiamen, Lin Hanzong's fate is tied to the seas. The sea has nurtured him and he has a deep love for the sea.

After graduation, Lin Hanzong was assigned to work in Beijing. Before leaving for Beijing, Lin Hanzong, not knowing when he might return, took a special trip to a seaside resort and savored several mouthfuls of the seawater of Xiamen. Lin's work in Beijing went very smoothly. Several years later, Lin Hanzong, after making some significant achievements, got a chance to return to Xiamen. Although his boss tried very hard to retain him, he decided to leave. Speaking of that time, Lin Hanzong said, "Had it been another place instead of Xiamen, I might have decided to stay in Beijing. But to Xiamen, I must go!"

His deep attachment to the sea ensured that Lin Hanzong worked on the seas after his return to Xiamen. From the management of Yundang Lake to Integrated Coastal Management (ICM), from leading the Marine Management Office to mediating various disputes, Lin Hanzong worked, although not a frontline engineer, behind the scenes as a mediator to promote Xiamen's ICM. Thanks to his efforts, Xiamen's ICM has made worldwide achievements.

※ From simple love to a lifelong mission

Throughout my life, I have never strayed far from the sea. I was born, grew up, and was educated in Xiamen. After I graduated from Xiamen University, I went to work in Beijing but returned several years later.

Because I grew up by the seaside, I am very attached to the sea. I still remember that when the decision was made arranging for me to work in Beijing, I hurried to the Seaside Resort of Xiamen University three days before I set off. I thought that I might never return, so I savored several mouthfuls of seawater as a tribute.

When I left Beijing to work in Xiamen, I chose a sea-related position, because I knew that sea was important for Xiamen's development, and that we needed to protect the sea and develop with moderation. Now, the seas near Xiamen are being developed at an even greater pace, which will boost Xiamen's economy. But moderation is a must, as many things cannot be repaired once they are broken.

Actually, my work in Beijing was also related to the sea. Back then, I was working on environmental protection. You could say that I was one of the first to work in environmental protection in China.

In Beijing in 1978, I started sea-related work. At that time, the Bohai Sea were heavily polluted, so the government set up a project called "Investigation into the Relationship between Pollution in the Bohai Sea and People's Health". I was a member of the project, marking the beginning of my sea-related work.

When I returned to Xiamen, my work was also related to the seas. One of the first sea-related projects I worked on was the rehabilitation of Yundang Lake. In 1982, when I returned to Xiamen, Yundang Lake had already become highly polluted. When I took up the mission and went to inspect the lake, all I saw was a stripe of dark and smelly water.

Back then, experts from around the country were invited to study the case and propose their suggestions and I summarized them into a treatment plan. After the plan was produced, it was approved by the Fujian Provincial Party Committee and the Provincial Government. The whole project was named "Short-and Medium-Term Treatment Project of Yundang Lake". When the Xiamen government presented the plan to the provincial authorities, we received six million RMB from the Discipline Inspection Commission of the Provincial Party Committee. Back then, Xiamen was very poor, and there was no money for building a sewage treatment plant. Our sewage pipes were installed on this money.

In the process, experts from Xiamen University and the Third Institute of Oceanography of the State Oceanic Administration suggested using natural resources such as seawater to wash away

Introduction to the speaker

Lin Hanzong, former deputy director of the Xiamen Municipal Environmental Protection Bureau and deputy director of the Xiamen Marine Management Office from 1996 to 2002, has been responsible for coordinating the work of the administrative group and the ExpertPanel of the Integrated Coastal Management.

the silt, while the Municipal Bureau of Parks and Woods supported using freshwater, as the supply of freshwater was very adequate in Xiamen at that time. I had visited many cities and noticed that most coastal cities were short of fresh water. Worried that Xiamen might face the same situation in the future, we were firmly against the use of freshwater. Municipal Bureau of Parks and Woods said that after seawater was introduced, "greenification" would become difficult, with pipes eroded. So I set out to convince them. I told them greenification would not be a problem, as long as we picked plant varieties wisely, as the plants in the coastal parks all grew very well. We would also find a scientific solution to protect pipes.

Using fresh or salt water was discussed, and after negotiations, it was decided that we would use sea water. However, how much sea water and the source brought another round of debate. After several discussions, the Expert Panel made the final decision. In this way, the pollution in Yundang Lake has been treated so successfully.

The pollution treatment of Yunlang Lake also helped ICM efforts. First, it is about reduction of sediment. Yundang Lake was originally a freshwater lake, so when sea water flowed in, sediments accumulated in the lake. By carefully controlling the amount of seawater going in, we made sure that the lake would not be flooded with sediments and that we did not need to wash the sediments away into the sea every year, which would also pollute the sea.

Second, treatment of pollution in Yundang Lake also helps the reduction of wastewater. There used to be a great deal of sewage in Yundang Lake. Sewage treatment plants were helpful but with limited effects. If the amount of sewage surpassed their capacity, the sea would become polluted. So, the pollution treatment helps improve the overall quality of sea water in Xiamen. Moreover, when Xiamen further develops coastal tourism in the future, Yundang Lake will be a component. Its improvement benefits the development of coastal tourism.

※ Setting up an agency to manage many different affairs

Residents in Xiamen have high expectations for cleaner seas. At that time, Xiamen Municipal People's Congress proclaimed that Xiamen must take advantage of its position as a port for further development. Since ancient times, Xiamen has always boasted its close relationship with the sea. As home to deep-water and silt-free ports, it boasts an advantageous position for foreign trade.

The government has always underscored marine development. In the 1990s, when an international organization carried out a project to study how to balance development of coasts with pollution prevention and control, Xiamen was picked as one of three case studies. After

▼ The Qianhui Lake, once a stain on the image of Xiamen, has become the most successful case of Xiamen Coastal Zone Integrated Management, also recognized internationally. (photo / Wang Huoyan)

that, we also studied how to develop the seas around Xiamen, how to take a multi-pronged approach and how to coordinate the work among different government agencies. From a management perspective, coordination was very important, so we set up a Marine Coordination Group, with an executive deputy mayor as its group leader.

The Marine Management Office was the leading branch for the group. In 1996, when Xiamen became the demonstration zone for an international research project, the Office was mainly in charge of ICM. Xiamen performed the best among the three zones picked, and took this chance to keep up with the world. Since then, 20 years have passed and Xiamen's model has demonstrated to the world how to achieve sustainable marine development by working on administrative and scientific fronts. This project has also made Xiamen more determined to continue this path.

In ICM, the Marine Management Office has played a very important role. Under the Chinese system, work could only be done with sound coordination. If the agencies involved went their own way, then nothing could be achieved. In this process, the Office has worked as a coordinator. It had many branches and those branches would meet regularly to discuss problems every month. In retrospect, setting up a coordinating agency has been very important for the success of Xiamen's ICM.

※ The Marine Management Office smoothes over management issues

At that time, I was also the leader of the Xiamen Environmental Protection Bureau. Working for the Bureau was intersected with many interesting stories.

In the early days, the Environmental Protection Bureau was responsible for ocean pollution monitoring, but couldn't handle it, because trash floating on the sea is difficult to deal with. Later, at the Marine Management Office, we decided

that land pollution was the responsibility of the Environmental Protection Bureau of Xiamen, whereas the Xiamen Bureau of Oceans and Fisheries took charge of sea pollution. The Marine Management Office monitors pollution, and then the actual sources of pollution are reported to its sister departments. In this way, the marine environment can be effectively maintained.

In the old days, whenever a typhoon arrived, trash would heap up on the shore, making the whole coastline very dirty. Later, it was handled by the Marine Management Office. A marine environmental sanitation department was set up by the Municipal Parks Bureau of Xiamen. They set up a team and commissioned a boat to deal with ocean trash. This work continues to this day.

1. Law Enforcement Officers were documenting facilities before removing fishing rafts.
2. Drones were launched to monitor island ecology and guard the sea.
3. Law enforcement at sea

Sometimes departments can be more effective when they cooperate. Take marine pollution accidents treatment for example. The Environmental Protection Bureau's research into the land sources of marine pollution could help us better grasp the overall problem. I remember that when a deputy mayor led a team to Xinglin Bay to inspect local development and construction, fishermen surrounded him on the boat and reported the death of their fish. An official immediately called me and asked the Marine Management Office for help. I took a quick look at the problem. First, the polluting factories that the fishermen reported were relatively far away from the aquaculture areas where the accident occurred. If the water was polluted by factory sewage, there would be dead fish along the route from the aquaculture area to the factory.

However, no dead fish were spotted close to the factory, therefore, the dead fish could not have been caused by factory pollution. Second, after we found out that the dead fish were all of a single species, experts from the Marine Management Office decided that this was a targeted infection rather than pollution. Through the coordination of various departments, the Marine Management Office clarified all the problems in the area, and the fishermen returned home satisfied. Think about it, if the Xiamen Municipal Bureau of Oceans and Fisheries handled the case alone, it would have approached other departments to conduct some investigations. This would be very troublesome and time-consuming. Therefore, the Marine Management Office can smooth things over when dealing with the pollution accidents by coordinating departments and division of labor.

Through the Marine Management Office, we will also coordinate with each other, making the whole Integrated Coastal Management more scientific and reasonable.

Take laws and regulations formulation for example. Administrative departments that formulate local laws and regulations in Xiamen often tend to focus on their own interests in the formulation process. With an institution like the Marine Management Office, regulations on coastal management can be formulated in a more comprehensive manner.

The Environmental Protection Bureau also has an ocean monitoring branch. Later, the Xiamen Municipal Bureau of Oceans and Fisheries and the Municipal Bureau of Parks and Woods set up ocean monitoring stations. Data gathered by different monitoring departments at different times and in different sites were different or even contradictory. Therefore, the Marine Management Office conducted coordination so that all monitoring departments could better cooperate and share data. If the data were contradictory, all departments should coordinate and figure out which data were accurate. Therefore, the Marine Management Office does play a significant role on multiple fronts.

I believe that the Office has achieved some major accomplishments in the past 20 years. The Office has helped publicize coastal management experiences of Xiamen to the world, developed Xiamen into a city planned around ports and established a science-based and reasonable functional positioning for coastal management in Xiamen.

In addition to administrative branches, this Office also has a Marine Expert Panel. Sometimes the Panel has divided opinions. However, different from administrative branches where conflicts arise due to conflicting interests, the Expert Panel has disagreements due to varied

◄ A cutter suction dredger working in the sea (photo / Wang Huoyan)

scientific and logical reasoning. For example, some experts on biological research suggested mangrove forests planted to preserve coastal zones near the Gaoqi International Airport. I expressed that mangroves were characteristic of increasing biomass. Additionally, mangroves attract many egrets and seabirds, and they would jeopardize airplane safety. Experts were persuaded, and gave up their proposal.

Experts generally consider problems from the angle of practical science, while as administrative officials, we tend to underscore the opinions of the municipal government. But when proposals submitted by the expert panel disagree with those of the government, we must take the whole picture into account.

※ Eastern and western sea areas: different locations, different requirements

Environmental protection means a lot to ICM. At that time, the Marine Management Office had three clear functions. The first was establishing scientific methods for marine development in accordance with environmental impact reports. For example, we need to be meticulous and attentive to functional orientation and potential impacts on tidal flow when it comes to land reclamation.

Efforts to implement ICM in Xiamen have been very successful, and we hope that the work can be improved in the future. Although we have already completed the preliminary international projects, we need to bear more results. Some management problems still remain. For example, in the past, some experts have resolutely opposed land reclamation. However, this view is narrow from my perspective. Therefore, I made more efforts in coordination.

When implementing ICM in the western sea areas, I argued that the area was paramount to developing marine economy. The key is to dredge the port, dig deep while still prioritizing environmental protection.

If the western sea areas are well protected, economic development is bound to ensue. But if someone proposed land reclamation in the western sea areas, I would not give way. On the contrary, I would be more open to land reclamation in the eastern sea areas. At that time, the tidal flows of this region were in good shape. At that time, I argued with our director on whether the sea area in front of the Conference and Exhibition Center should be reclaimed. The Center was originally planned to be located in the countryside, far away from its final location, which would hinder development due to inconvenient transportation. Reclamation was called for after finalizing the present site. At that time, many experts held that

land reclamation would cause pollution. They put forward two arguments. One was that the resulting tidal changes might negatively affect nearby Huangcuo beach. And many people in Jinmen were against reclamation because they thought it would cause sediments to be carried to Jinmen, affecting their sea area. These two problems were real, but I still supported this proposal. After we inspected many locations, it was the only location suitable for the Exhibition Center. But in the long run, with cross-strait relations improving, Xiamen and Jinmen can cooperate to treat sedimentation. Later, I suggested building a cofferdam before land reclamation, and the builders agreed.

In fact, in the Integrated Coastal Management of Xiamen, all work keeps pace with the times. For example, now, during coastal development, we need to consider how best to contain floodwaters when the flood comes. In the past, we only considered pollution, and never thought of flood discharge. This made us encounter many setbacks. Now, we have to learn from past lessons.

※ Looking at problems from a historical perspective

Xiamen is currently developing Tong'an Bay, and it involves a great deal of reclamation. In my opinion, in order to further develop transportation, it is necessary to conduct land reclamation. But I suggest that we should not do that without a plan. In some areas, land reclamation is easy, but afterwards, the tidal flows will be affected. It is advisable to avoid sedimentation as much as possible and make sure tides flow. Besides, tidal influx should remain unchanged after reclamation. Therefore, in this case, we also call for experts' opinions.

▲ A new and beautiful residential complex by Xinglin Bay

Land reclamation remains contentious in ICM. In the past, the Expert Panel often disagreed with the administrative group on this matter and I need to coordinate in this process. For example, in the western sea areas, I support the expert opinion that reclamation should proceed only after very careful and strict review. While in the eastern sea areas, I support the administrative group. The experts stuck to their opinions. But after all, administrative departments should consider all kinds of problems in the early stages and conduct coordination. Take the Exhibition Center mentioned above for example. Many experts opposed reclamation, but in the end, the reclamation did not exert many negative effects.

Sometimes, however, compromise should be made between experts and officials. For example, in Tong'an Bay, from a science-based perspective, the reclamation area should not be immense, but for economic development, reclamation was necessary. Another typical case is Xinglin Bay. Some different opinions still existed. Experts suggested that although the area was already silted and could be filled, but after reclamation, the rate of siltation might accelerate further. But concerning urban development at that time, land reclamation was mandatory.

In retrospect, some people may say that our previous work was insufficient and not environmental-friendly. I think that view is wrong, because in the early days, most people did not understand the concept of environmental protection, and all the plans were made for economic ends. Things should be seen from not only a science-based, but also a historical perspective. I often tell others to have this kind of perspective, and analyze both the merits and demerits of our decisions, without jumping to conclusions simply by today's standards.

Strong Science and Technology Bases for Integrated Management

Hong Huasheng

As the first female doctor of oceanographic science to return to China from overseas study, Hong Huasheng has been recognized as one of the leading oceanographers at Xiamen University. Over the past two decades, a team of scientists led by Hong Huasheng, have offered strong science and reliable technological support for the Integrated Coastal Management in Xiamen.

"All work must be forward-looking and be guided by science. " In interviews, Hong Huasheng often emphasizes this principle, which has also been her approach to years of research on the oceans. Thanks to the efforts of scientists like Hong Huasheng, Xiamen has successfully formulated the Marine Functional Zoning and has become a model both at home and abroad.

Besides her work on Integrated Coastal Management in Xiamen, Hong Huasheng has also spared no effort in raising public awareness about marine environment. Thanks to her efforts, the World Ocean Week in Xiamen, a platform for international exchange, has greatly contributed to the sustainable development of coastal zones by increasing awareness of integrated management efforts.

The ultimate goal of scientific research is to improve people's lives. Livelihood is what concerns Hong Huasheng most. "I'm a scientist, and it's my duty and responsibility to make contributions to the development of the marine economy, the protection of the marine environment, and to enhance people's livelihoods. " This is Prof. Hong's firm belief about her beloved marine science.

※ Set up a Marine Expert Panel

The Integrated Coastal Management in Xiamen reached a turning point in 1994 when rehabilitation began on Yundang Lake. How did this program begin?

In 1992, the United Nations Conference on Environment and Development passed Agenda 21, which devoted a whole chapter to the sustainable development of oceans, and mentioned Integrated Coastal Management. It was a rather new concept in China back then. An international project on the integrated management of oceans was designed to be piloted somewhere on the East Asian seas. Xiamen was chosen when this project came to China. On the recommendation and with the support of the State Oceanic Administration, the Xiamen Demonstration Project for Prevention and Management of Marine Pollution in East Asia seas was co-launched by the Global Environment Fund, the United Nations Development Program, and the International Maritime Organization, which was the starting point for Integrated Coastal Management in Xiamen.

As a transport hub in East Asia on the western coast of the Taiwan Strait, since ancient times, Xiamen has been an important port of trade connecting coastal regions in southeastern China to the outside world. Since the inception of Reform and Opening up, Xiamen has become an iconic port city with famous natural scenery, and one of five special economic zones in China. Oceans and coasts are the most important economic and biological resources and lifeline for Xiamen, and they have played a critical role in the city's economy. The fast economic growth in the 1980s and 1990s once put enormous pressure on Xiamen's coastal environment. Only by protecting oceans and coast properly can Xiamen grow sustainably. As a result, how to best use coastal marine resources and tackle environmental problems has become a headache for Xiamen's sustainable development. Luckily, Xiamen's senior government officials then were open to new ideas and hoped to change the old ways of coastal management in Xiamen.

Awareness and international projects and concepts alone are not enough. What turns theory into reality is practical application. We could not copy other countries without second thought in terms of Integrated Coastal Management.

New management methods only come about when advanced science and new technology have been put in place. At that time the Disciplines of Oceanography in Xiamen University had already become well-established, but the problem was that researchers hardly cooperated with government agencies. When the Xiamen municipal government began implementing Integrated Coastal Management, the government decided to set up a Marine Expert Panel and invited me to head this group. I didn't agree at first, because I was a scientist and knew very little about public affairs. But later I was told that my background in oceanography could strongly support the implementation of Integrated Coastal Management. I decided to get on board as I was hopeful that the program could help protect the seas near Xiamen.

The municipal government had set up several working groups of experts, including a group for civil engineering a group for science and technology. I suggested we needed only one high level expert group made up of authoritative members, which could spearhead coastal management specifically. Another suggestion was that, besides university professors, a certain proportion of members must be government and administrative experts. Based on these two suggestions, a special marine expert group was set up and, different from today's working groups, included experts with multiple-disciplinary background.

※ Develop and ImplementMarine Functional Zoning

The first thing we needed to do upon setting up the Marine Expert Panel was to figure out the proper way to delineate marine functional zones in Xiamen.

The marine environment has multiple layers, and it is inevitable to see a variety of industries develop in a one sea area, which is likely to create conflicts between the different participants as they engage in the development of shared marine resources. There were conflicts between shipping, fishing and tourism industries, and between sea reclamation and marine conservation. All these conflicts could result in unregulated and excessive development, which in turn could lead to depletion of marine resources, worsening pollution, and even the collapse of the marine ecosystem.

Marine Functional Zoning avoids blind, unregulated, and excessive development, and contributes to wise decision-making and sustainable utilization of marine resources. It is therefore a cost-effective method to tackle conflicts between different uses.

Introduction to the speaker

Hong Huasheng, Distinguished Professor of Xiamen University, Honorary Director of the State Key Laboratory of Marine Environmental Science (Xiamen University), chief scientist of the Coastal and Ocean Management Institute (COMI) of Xiamen University. In 1995, she served as head of Xiamen Marine Expert Panel and deputy director of Xiamen Coastal Zone Sustainable Development Training Center. She has participated in and led the development of Xiamen Marine Functional Zoning and the sustainable development of Integrated Coastal Management.

▼ After Xiamen marine functional zoning, Southeast International Shipping Center located its headquarter in Xiamen Haishu, which is the fourth international shipping center in China after Shanghai, Tianjin and Dalian. (photo / Wang Huoyan)

Marine Functional Zoning is defined as a decision-making process to formulate effective mechanisms and solutions to optimize the distribution of resources. Its main purpose is to regulate human activities, so as to minimize disturbances to the marine environment and while satisfying people's demand for consumption and production demand through a functioning marine ecology.

However, the marine economy is a large-scale interdisciplinary system that involves multiple regions and industries, which since the early days has made the management and coordination of marine areas more complex than land. In addition, all the government agencies involved in coastal management in China acted according to the powers granted by relevant laws and regulations of their respective branches. Their respective needs were their only concern when they formulated and implemented marine development plans. The lack of coordination and testing mechanisms inevitably led to chaotic development and conflicts between different uses in the shared area, which resulted in the waste of financial and labor resources. The underlying cause of these problems was flawed legislation on coastal management. Therefore, it was also very important to establish sound laws and regulations related to Marine Functional Zoning, equivalent to those for land management.

The Marine Expert Panel and sea-related agencies held ten meetings to formulate a detailed plan for marine functional zoning. At the last meeting, the then Xiamen mayor was present, and after we reported the plan to him, he immediately informed us of the approval of the plan by the Xiamen government. In 1997, the plan was passed by the municipal People's Congress, which gave it legal authority. Looking back on it 20 years later, we are glad to see our concerns back then were well addressed in the plan.

※ West Sea Areas: Give place to port industry

Marine zoning basically delineates the proper usage of marine resources. Xiamen was the first city in China to begin marine functional zoning. In the early stages, development of the seas was chaotic. For example, both fishing and shipping were conducted in one area, creating many conflicts. It was urgent to begin Integrated Coastal Management to improve the functioning of sea areas.

In order to formulate plans for Marine Functional Zoning, together with some experts, I went on a research tour of the current sea usage in Xiamen, and we paid special attention to the coastlines and ocean currents, before we conducted further study on issues related to the usage of sea resources, such as the impacts of vast sea reclamation. The research team carried on with the issues concerning the future development of Xiamen, and how seas

▲ Crowded cages under Haicang Bridge in the old days

of Xiamen should be used. The Marine Expert Panel established the framework, which defined the use high priority, maximum benefit for multi-resource uses, and restricted functions of each sea area.

We started work in the western sea areas first. The traditional industry in the western sea areas is aquaculture. At that time, there was only a small Port at Dongdu, and Haicang Port had not yet been built. On flights to Xiamen, foreign guests might be confused to see white patches dotting the western sea areas. In fact, those were patches of foam farmers laid in the sea for fishing. At that time, the port industry in Xiamen was emerging as an economic pillar. However, transportation channels in the entire Dongdu area were clogged by aquaculture. When entering Dongdu Port, ships were often ensnared by fishing nets.

Consequently, it was important to answer the defining question: what was the major purpose of the western sea areas? Despite the large number of people engaged in aquaculture, we suggested the port industry should be prioritized in the western sea areas, since the Xiamen government had put forward the vision of "building an international port city with beautiful scenery". Accordingly, only by giving ports top priority over any other activities in the western sea areas could the port industry grow in Xiamen. Thanks to this vision, Xiamen has grown into an international shipping hub.

Ports serve as the dominant function of the west sea areas, with tourism serving as a compatible function. Dayu Island was already under development at that time. The city built mangrove forests there and the island became home to egrets, and the scenery was gorgeous. Because a waste water treatment plant was located in the western sea areas before our planning began, discharging of waste water was also included as a compatible function in the early version of Marine Functional Zoning. Later, this function was gradually phased out.

Aquaculture was the industry that required the most urgent restriction in the western sea areas. We worked hard to remove it. At that time, sea-related government agencies in Xiamen reached a consensus that the port industry in the western sea areas should be developed, but disagreed on the necessity of phasing out aquaculture. Experts from the Marine Expert Panel tried their best to persuade the leaders. We raised two main arguments. The first was that aquaculture was likely to impede the construction of ports and navigation in the western sea areas. The second was that aquaculture would lead to the deterioration of water quality, leading to the industry's own destruction. Aquaculture required Type-2 water, while shipping, which only required Type-4 water, was quickly growing. When the quality

1. Today's busy shipping under Haicang Bridge (Photo / Yonghe)
2. The view of silted river from the perspective of Haicang Bridge in the past
3. Today, this area is hailed as Haicang's the Bund. (photo / Zheng Weiming)

of water worsened as the shipping industry developed, the aquaculture industry would collapse.

※ Eastern Sea Areas: Weigh the advantages and disadvantages of reclamation projects

After the work in the western sea areas, we turned our attention to the east an sea areas. At that time, tourism was the dominant function of the eastern sea ares. One problem emerged when we were doing the feasibility study for sea reclamation in the area.

Sea reclamation near the Convention and Exhibition Center aroused particular controversy. Officials from environmental protection agencies strongly opposed it. So the Marine Expert Panel organized an environmental impact assessment of the area. We found that the area was a piece of eroded land rather than sandy beaches, so reclamation of the land would have little impact on the functioning of marine ecology. We could see the pros and cons of

1. On April 23, 2011, 5,000 mangrove saplings were planted at Xiangtan Lower Wetland Park in Xiang'an. (Photo / Dai Shujing)
2. An aerial photo of the mangrove forest at Tamitan Wetland Park. (photo / Zhu Yili)

sea reclamation. The key was which mattered more. Back then there was an urgent need to develop the convention and exhibition industry in Xiamen, which would occupy large land area. After weighing the pros and cons, we considered sea reclamation project of this area was feasbile. It was because of the reclaimed two square kilometers that we have a flourishing convention and exhibition area in Xiamen.

Although the area around the Convention Center was suitable for reclamation, we determined that Wuyuan Bay, which was originally planned for land reclamation, should never undergo reclamation. I demonstrated that one important standard of international port cities with beautiful scenery is the number of yachts. If marine tourism was to grow in this city, we had to develop the yacht industry. As income increases, people no longer wish to travel by car, but prefer yachts. Twenty years has passed since we proposed this idea. In retrospect, our proposal makes sense and provides a good foundation for the development of yacht industry in Xiamen. If Wuyuan Bay had been filled in at that time, there would not be a prosperous marine tourism area there today. Although the area of Wuyuan Bay is much

smaller, at least the most important area has been preserved.

For construction of the Wuyuan Bridge, the Marine Expert Panel also made some suggestions, in particular, for the shape of the bridge. We suggested building a bridge with one big arch; otherwise it would be impossible for yachts to pass. There were no yachts in Xiamen at that time, and many people simply could not understand the importance of the arch, so they did not understand the design of the bridge. However, we should be forward-looking in decision-making.

Another subject for the Marine Expert Panel was the Haicang port. The group believed that the entire western region was relatively narrow, and would limit the growth potential of the Dongdu Port. Therefore, Haicang, including Zhangzhou, as a would-be international port, should be the focus of development. I proposed that the coastline of Haicang should be developed into a port. Now in retrospect, without implementing coastline protection at that time, there wouldn't be a booming port industry in Haicang, let alone the Southeast International Shipping Center in Xiamen. Therefore, the proper functional zoning of the sea will lay a solid foundation for the marine economic development of Xiamen and surrounding areas.

Another focus of our efforts was the Wutong Ferry. At that time, it was proposed that real estate should be developed in the Wutong Ferry coastal area. The MarineExpert Panel disagreed and advised that it was best to preserve the coastline of Wutong Ferry, because it was likely to become an important hub connecting Kinmen in the future. But some leaders were skeptical, because 20 years ago, we had no formal contact with Kinmen at all, but I thought we would see changes very soon. As a result, the coastlines of Wutong Ferry has been preserved because of our suggestion.

My stance had never changed when it comes to the marine functional zoning of various ports and harbors—There were not many natural coastlines in Xiamen at that time, and the remaining valuable coastlines had to be preserved.

※ Scientific evidences for breaking and coffering the dams

Immediately after marine functional zoning was completed, we started to develop a Geographic Information System (GIS) with a scale of 1:5,000. The GIS System has enabled the effective operation of the Marine Function Zoning. The Coastal Management Information System, which is continuously updated with the utilization of the high-precision global positioning system (GPS) and application of the satellite remote sensing (RS) technologies. These technologies have helped constantly improve the precision and updating of the zoning scheme.

When we begin to exploit sea resources within the zoning areas, we should know the size of the used area before making any decision; therefore, we need to know the information in advance through the information system. In addition, users of sea resources must pay a certain amount of fees. How much should they pay? This was also a task of the Marine Expert Panel to help set up a sea use permit system. The proposals of the Marine Expert Panel were unheard-of in Xiamen, from Marine Functional Zoning to the Data Management System, and to the paid use of the seas. Later, national legislation on marine issues also borrowed a lot from the experiences in Xiamen. Xiamen has been a pioneer in Integrated Coastal Management, and through its successful efforts, has become a model for East Asian countries.

The work on Integrated Coastal Management has helped me realize the importance of promoting a better understanding of the issues, in particular, for government officials. So we, together with the State Oceanic Administration and authorities in Xiamen, established the International Training Center of coastal sustainable development, and offered training for officials from both China and other East Asian countries. Later, many senior government officials attended lectures there, giving them a better understanding of the Integrated Coastal Management.

In the first phase of Xiamen Integrated Coastal Management, the Marine Expert Panel completed the Marine Functional Zoning, and in the second phase of Xiamen Integrated Coastal Management, focused on marine ecological rehabilitation. Two major tasks of ecological rehabilitation were the breaking of Gaoqi Dam and Maluan Bay Dam. Was it necessary to break the dams? If so, and how to do it? The Marine Expert Panel's opinions were needed. Take the Gaoqi Dam as an example. At that time, more than 10 meters of silt had accumulated at the bottom of the Gaoqi Dam, and the entire sea had lost 60% of its flushing ability. Similar to flushing a toilet, if the sea current was too slow, many of its natural functions would gradually die out. Like Yundang Lake, where the hydro power generation was weak, silt accumulated at the bottom of the lake. Waterways surrounding Songyu Island and Gulangyu Island also had serious siltation problems. The Jiulong River flowed into the seas of Xiamen and brought in mud, which was eventually deposited in the seas of Xiamen. The Marine Expert Panel believed it necessary to break the dams after inspection. The question was how to best go about it. It was a complicated project, and it was difficult to reach consensus. So we proposed to develop a model first. Together with researches from Xiamen University and the Third Institute of Oceanography of the State Oceanic Administration, we developed a model. Then, through on-site testing and scientific demonstrations, the Panel proposed another plan for the Gaoqi Dam. Rather than a full

▲ *Yearning* a sculpture (photo / Wang Huoyan)

blasting, a gap 800 to 1,000 meters in length should have the same effect. Again, scientific research and demonstrations were needed when it came to the height of the cofferdam for Maluan Bay.

These restoration projects have dramatically improved the ecological environment of Xiamen sea waters, maintained the marine ecological balance, restored marine resources, and improved the marine landscape, thus boosting port development, shipping transportation, coastal tourism, coastal industries, and other emerging high-tech marine industries.

In conclusion, Integrated Coastal Management must be based on thorough scientific research. This requires the government to have frequent contact with research departments, allowing them to put forward their own opinions and concerns before making decisions. When doing marine projects, we must do thorough research and provide ground for any potential plan. Marine projects are different from land projects, whose environmental impact assessment is limited to one specific area, while marine projects involve a wide range of issues and its environmental impact assessment is comprehensive and inclusive.

※ Communicate marine monitoring data and information

We have made great efforts to promote Xiamen's Integrated Coastal Management on the international stage; Xiamen has held the International World Ocean Week annually since 2005. In 2004, the International Maritime Organization and the Ministry of Environmental Protection co-hosted an international environmental conference in Xiamen. One important issue raised was related to oceans. Xiamen's Integrated Coastal Management by then had made some significant achievements. A UNEP offcial attending the conference was very impressed by Xiamen, and she highly admired Xiamen's model and suggested that this model could be promoted in other countries, not only in East Asia, but also in other countries around the world.

At that time, Stockholm held a World Water Week, which was quite a success. So that the official proposed that since the Stockholm water week was primarily concerned with fresh water, Xiamen could hold an ocean water week. This proposal quickly received support from the Xiamen government. Therefore, the "Xiamen International Ocean City Forum" was launched in 2005, which was the predecessor of the Xiamen World Ocean Week (WOW).

An important issue of ICM is data and information sharing. Take marine monitoring as an example. the Ocean and Fishery Bureau (OFB), the Environmental Protection Bureau (EPB) and the Water Conservancy Department (WCD) all conduct ocean monitoring separately, and although it is important for them to share information, they often fail to do so. Previously, data on water quality from different agencies could differ radically. In fact, water monitoring was dynamic and the conclusions in different seasons could be completely different. I suggest that Xiamen should further improve its data exchange platform among all-related agencies. A sustained and integrated environmental monitoring system is envisioned for the future as an integrated information system that systematically acquires and shares data and information to serve policy, management, and other specific needs of many user groups (government agencies, industries, scientists, educators, NGOs, and the public).

The Xiamen's ICM experience in the utilization of scientific research results and innovative technologies (e.g., Marine Functional Zoning, GIS Information Management System, Numerical Modeling, Integrated Environmental Monitoring, etc.) is a showcase on how strong science and advanced technology could help to implement Integrated Coastal Management for ocean and coastal substantial development.

Marine Expert Panel Provides Robust Technical Support

Xue Xiongzhi

Xue Xiongzhi, Executive Dean of Institute of the Ocean and Coastal Development, Xiamen University, served as head of the secretariat of the Marine Expert Panel of Xiamen Integrated Coastal Management in 1995. He is now a member and Secretary General of Xiamen Marine Expert Panel. He has participated in Xiamen Marine Functional Zoning and other major projects of coastal management. *Xiamen Integrated Coastal Management in the Past Ten Years* is one of his major works.

Radiant sunlight, rippling waves, and golden beaches define Xiamen and give it a touch of romance. In this glamorous city, blue skies and white clouds frame the beaches, and all who come here are mesmerized by her charm.
Xue Xiongzhi is a witness to the tremendous changes that have taken place in this city over the past two decades.

The underlying reason for this charm is the distinctive Integrated Coastal Management (ICM) in Xiamen. In 1994, Xiamen took the lead in introducing this concept. In 1995, with the formation of the Marine Expert Panel in Xiamen, Xue Xiongzhi, as the leader of this secretarial Panel, engaged in Xiamen Integrated

Coastal Management. Xue Xiongzhi always thought highly of the projects such as Functional Zoning, opening up of causeways, and marine ecological rehabilitation. Technical expertise underpinning the Expert Panel is an important backup for the ICM in Xiamen.

Corroboration by experts, cross-sectorial coordination and cross-disciplinary integration have made Baicheng Beach, Wuyuan Bay and the Huandao Road more and more popular. In his words, the support of the Marine Expert Panel is not only about technology and engineering, but also fundamentally about policymaking. From his point of view, Integrated Coastal Management is dedicated to tackling disputes over the proper use of resources by forging policies and strategies in a sustained and active effort to control the environmental impact of human activity on the coastal zone. This has become an effective instrument for grappling with the negative effects of resource exploitation on the environment.

Xue Xiongzhi holds dearly the sea close to Xiamen University, where he has worked and lived for over 20 years. In the future, the management efforts will persist and the situation will become more amiable. He often enjoys the area by taking walks along the beach and the Huandao Road. He believes that the waters of this area will become even more crystal-clear in the future.

※ Government departments interact well with the Expert Panel

Both in the past and at present, Integrated Coastal Management in Xiamen has been at the forefront of parallel efforts across the country. The concept of ICM emerged on the international stage in 1992, and in 1994, Xiamen introduced this concept and further promoted this initiative. Over the past two decades, Xiamen has made good headway in functional zoning, passing legislation, including the *Usage and Management Regulations on Coastal Areas*, and ecological rehabilitation.

The first round of IntegratedCoastal management, which lasted 10 years, focused on introducing this new concept and establishing a mechanism, with special emphasis on pollution prevention and treatment along the coast.

After 10 years of strenuous efforts, practices have demonstrated that ecological rehabilitation paid off with ecological, social and economic benefits. These efforts mainly cover the following aspects: marine ecological environmental rehabilitation and marine biological resources rehabilitation, including breeding programs and reintroduction, mangrove ecological rehabilitation, exotic invasive species elimination, and marine rare species conservation. Progress has brought ecological, social and economic dividends.

Ecology comes first. I had the first-hand experience with the exploration and establishment of the new sustainable development paradigm in Xiamen. In 1995, propelled forward by the Xiamen Demonstration Plan of East Asia Marine Pollution Prevention and Management, the Xiamen municipal government set up a Leading Group for coastal management and coordination headed by Mr. Zhu Yayan, the then Executive Vice Mayor. The group members included leaders in planning, economics, technology, environmental protection and aquaculture. This group has an office and the Marine Expert Panel is composed of renowned experts and professors in marine studies. I was

Introduction to the speaker

Xue Xiongzhi, Executive Dean of Institute of the Ocean and Coastal Development, Xiamen University, served as head of the secretariat of the Expert Panel of Xiamen Integrated Coastal Management in 1995. He is now a member and Secretary General of Xiamen Marine Expert Panel. He has participated in Xiamen Marine Functional Zoning and other major projects of Ocean Management. *Xiamen Integrated Coastal Management in the Past Ten Years* is one of his major works.

quite young at that time and mainly worked as a liaison between the Expert Panel and the municipal government.

The effective implementation of this initiative was inseparable from the participation of the Expert Panel, formulation of *the Marine Functional Zoning*, management, and sustained technical support. I recall that we had a meeting with leaders from the municipal government at a 2-month interval. Of course, this is related to the attention placed on the issue by the then Vice Mayor Zhu Yayan . At the meeting, experts would give feedback on hotspot marine issues to their superiors and they would exchange views with experts and share their vision on how to offer assistance. It is fair to say that this Expert Panel acting hand in hand with the Marine

◄ Oceans are the most important resource, and arine economy is one of the most important economic pillars. Sailing is very popular in Wuyuan Bay, which has held many international sailing events. (Photo / Wang Huoyan)

Management Office truly embodies the scientific operation of Xiamen's ICM and has yielded remarkable results.

We have made pioneering breakthroughs. Marine Functional Zoning in Xiamen had never been done before, so we started it, and scaled it up. In fact, this should have begun long ago. But without a coordination mechanism, implementation was impossible. The zoning was delayed because no parties concerned were willing to give away their interests. The disputes in the western sea areas were quite acute at the time. Issues like aquaculture, tourism, pollution treatment, sea navigation, Chinese White Dolphin and egret sanctuaries all complicated the situation. Previously, the harbor was situated in the east. When ships entered the harbor, they had to pass the channel in the western sea areas. Unmanaged aquaculture made it difficult for ships to go in. After that, the Expert Panel took over the project. A firm stance and the support from Executive Vice Mayor Mr. Zhu and Mayor Mr. Hong Yongsheng at that time were the key to addressing the zoning issue through over 5 Mayoral Work Meetings and 10 revisions. The Marine Functional

1. A group photo of experts group on the 20th Anniversary of Xiamen Coastal Zone Integrated Management
2. Experts Forum in the old days
3. Heated discussions at a meeting for re-election of members

Zoning stipulated that the dominant function of this sea area was navigation, intended for sea route and wharf transportation. Other industries should be only secondary. In this way, aquaculture was restricted with no room for further expansion. And the previous haphazard utilization of sea routes was resolved.

Functional zoning is a necessity, as it divides regions with specific functions, by taking into account the diverse development models, geological location, natural resource and environmental conditions, and socioeconomic requirements. So it is fair to speak that the Marine Functional Zoning in Xiamen is an integral part of Integrated Coastal Management.

I remembered explicitly, that while compiling *Xiamen Marine Functional Zoning*, the Expert Panel led the compilation group and coordinated with different departments, building on marine research outcomes and development progress over many years. The document was supplemented with a functional zoning map at 1:5000-scale, leapfrogging from the prior small-scale measurements and applying the GIS. Regulators can be more successful in the managing marine zoning with the shrewd usage of scientific, endemic and operational features based on this outcome.

On October 22, 1996, the *Xiamen Marine Functional Zoning* was passed at the Executive Meeting of the municipal government. From 1996 to 2010, the zoning plan was designated as a "critical foundation" for marine protection efforts, regional development and legislation, such as the regulations on Xiamen coastal management.

Over the past 20 years, the Zoning plan generally went through one revision every 5 years due to changes in urban construction requirements and the environment. The two revision sessions I participated in did not result in big changes. The revision resolved issues like the lack of communication between the experts and government officials and disconnection between science and management. It also had a solid scientific foundation and robust implementation. Science and technology, via the Expert Panel, plays an instrumental role in corroborating the feasibility of plans and enhancing officials' knowledge. The Expert Panel constitutes the preeminent scientific body of the municipal government for ICM. Panel members have actively engaged in policy making and made many contributions.

※ The Expert Panel provides technical support for policymaking

When a car passes over the Dadeng Bridge, the great bridge that connects Xiang'an to Xiamen, swaths of cultivated shoals and salt-fields can no longer be seen. However, the profile and status of the Dadeng Island has been enhanced thanks to land reclamation efforts. Xiamen's second airport has been constructed here. A new gateway to the world has been opened.

The Expert Panel was integral to the completion of the Zoning project. It is fair to say that integrated management has reached a pinnacle from this angle. Since 2002 when the breeding and reintroduction of rare marine animals started, the efforts have proceeded smoothly. From my perspective, the Expert Panel played an essential role in providing scientific and technical support more to policymaking than to engineering.

The construction of the Huandao Road also exemplifies the technical support provided by the Expert Panel. The philosophy of the road's design was "catching more sight of the sea upon coming closer to the sea". To achieve this goal, without the advice of the

Expert Panel, original vegetation in this area, essentially a protected forest, was cut down to transform the area into a beach. This aimed to allow people to enjoy the sea. But one issue emerged—who was entitled to this convenience? To be frank, this initiative mainly benefited drivers instead of tourists. However, sufficient beaches, forests, and shaded areas should be preserved to retain tourists.

In the midst of project evaluation and verification, we found that despite its glamor, it was hard to retain tourists due to the scorching sunshine. Subsequently, the Expert Panel suggested that we needed more shaded areas and trees. As a result, the Huandao Road was lined by more trees, but they were still not adequate. Then we proposed this same suggestion once again for the east Huandao Road and took action accordingly. The trees along this road are more like a belt of coastline than of the sea. When tourists cross this lane, it is like walking in the shade of a forest.

Although the Huandao Road is built against beautiful landscape, the destruction of natural vegetation during its construction has left many regrets. In nature, when sand is left to its own devices, after several decades, it may submerge the entire roadbed. Because natural beaches emerge according to their own rules, they require ample room. Right now, some parts have been destroyed and in event of more sandstones accumulation, it is bound to sabotage the whole roadbed.

The Expert Panel has done a superb job in providing technical support. It has worked well to conduct evaluations in regards to handling some controversial projects that might have resulted in marine destruction.

According to the old-fashioned development thinking, Wuyuan Bay might have experienced a larger-scale land reclamation. Upon hearing these proposals, our Expert Panel quickly denied them during their evaluation. To conserve its ecology, we suggested that the coastal region should be maintained, Zhongzhai Bay opened up, and the middle part widened and deepened. The subsequent popularity of Wuyuan Bay as a resort area is attributable to this decision.

The "Wuyuan's sailing beauty" has entered the lexicon and is synonymous with the area's most magnificent landscape, bringing to mind a spectacular scene. Thanks to its high-quality water, fresh air, an open landscape and gentle breeze , Wuyuan Bay is acclaimed as an exceptional area for sailing by those in the sailing circle. It also enjoys fame as the most beautiful sailboat harbor, opening Xiamen up to the rest of the world.

Xiamen is a city adorned by flowers and the blue sea. It has been awarded a series of honors such as "National Garden City", "National Environmental Protection Demonstration City", and "International Garden City". Xiamen, as a longstanding advocate for ecological conservation, endeavors to beef up ecological rehabilitation efforts on its coastal belt. This work became one of the most significant tasks of ICM during the second ten-year period, which was undertaken by the Marine Department. Vegetation rehabilitation has also made tangible results.

At the same time, we have combined coastal ecological conservation with tourism, forming a virtuous cycle. All these are landmark results of ecological rehabilitation. Opening up the causeway is often regarded as the most typical and effective project of its type. It is intended to address the largest environmental capacity issue in the entire western coastal region and Tong'an Bay. Because after the opening up of the causeway, the coastal region could then be channeled and its water exchange and pollution holding capacities could be scaled up. These play a cardinal role in ecological rehabilitation.

However, the project remained contentious and took over 10 years to be executed. The disputes mainly covered two aspects. The first one is cost, involving unforeseen problems with pipelines and electric grids in addition to the causeway itself. If they were not redistributed and laid out properly, then the water and electricity supply might be cut off once the causeway was opened up. On the other hand, some veterans deemed the causeway as a symbol of political action and the result of the forefront position of Xiamen at that time. Many people strongly opposed the opening of the causeway. But from my perspective, we can adopt other measures to commemorate this part of history. Because of technological improvements, not only causeways can connect the area to the continent; we also have bridges and tunnels. On the whole, these two points of contentionn delayed the progress on this front for 10 years.

I remembered clearly that at that time Mayor Hong Yongshi went on a field trip to Xiamen and asked us why the sea in Xiamen could not be as clear and blue as seas in other places.

Admittedly, the reasons are multi-faceted. First of all, the coastal region was semi-enclosed. Turbid sea waters were surely associated with bad environmental health. Besides, it also was related to land reclamation. Land reclamation in the past was not very stringently controlled. Taking the Zhaoyin development zone in Zhangzhou as an example, when they reclaimed land in the area, they used explosives to fill the area with mountain soil. Just imagine how much silt flew into the sea. If the silt was solid, it might

Introduction to the speaker

Lin Hanzong, former deputy director of the Xiamen Municipal Environmental Protection Bureau and deputy director of the Xiamen Marine Management Office from 1996 to 2002, has been responsible for coordinating the work of the administrative group and the ExpertPanel of the Integrated Coastal Management.

the silt, while the Municipal Bureau of Parks and Woods supported using freshwater, as the supply of freshwater was very adequate in Xiamen at that time. I had visited many cities and noticed that most coastal cities were short of fresh water. Worried that Xiamen might face the same situation in the future, we were firmly against the use of freshwater. Municipal Bureau of Parks and Woods said that after seawater was introduced, "greenification" would become difficult, with pipes eroded. So I set out to convince them. I told them greenification would not be a problem, as long as we picked plant varieties wisely, as the plants in the coastal parks all grew very well. We would also find a scientific solution to protect pipes.

Using fresh or salt water was discussed, and after negotiations, it was decided that we would use sea water. However, how much sea water and the source brought another round of debate. After several discussions, the Expert Panel made the final decision. In this way, the pollution in Yundang Lake has been treated so successfully.

The pollution treatment of Yunlang Lake also helped ICM efforts. First, it is about reduction of sediment. Yundang Lake was originally a freshwater lake, so when sea water flowed in, sediments accumulated in the lake. By carefully controlling the amount of seawater going in, we made sure that the lake would not be flooded with sediments and that we did not need to wash the sediments away into the sea every year, which would also pollute the sea.

Second, treatment of pollution in Yundang Lake also helps the reduction of wastewater. There used to be a great deal of sewage in Yundang Lake. Sewage treatment plants were helpful but with limited effects. If the amount of sewage surpassed their capacity, the sea would become polluted. So, the pollution treatment helps improve the overall quality of sea water in Xiamen. Moreover, when Xiamen further develops coastal tourism in the future, Yundang Lake will be a component. Its improvement benefits the development of coastal tourism.

※ Setting up an agency to manage many different affairs

Residents in Xiamen have high expectations for cleaner seas. At that time, Xiamen Municipal People's Congress proclaimed that Xiamen must take advantage of its position as a port for further development. Since ancient times, Xiamen has always boasted its close relationship with the sea. As home to deep-water and silt-free ports, it boasts an advantageous position for foreign trade.

The government has always underscored marine development. In the 1990s, when an international organization carried out a project to study how to balance development of coasts with pollution prevention and control, Xiamen was picked as one of three case studies. After

▼ The Qianhui Lake, once a stain on the image of Xiamen, has become the most successful case of Xiamen Coastal Zone Integrated Management, also recognized internationally. (photo / Wang Huoyan)

that, we also studied how to develop the seas around Xiamen, how to take a multi-pronged approach and how to coordinate the work among different government agencies. From a management perspective, coordination was very important, so we set up a Marine Coordination Group, with an executive deputy mayor as its group leader.

The Marine Management Office was the leading branch for the group. In 1996, when Xiamen became the demonstration zone for an international research project, the Office was mainly in charge of ICM. Xiamen performed the best among the three zones picked, and took this chance to keep up with the world. Since then, 20 years have passed and Xiamen's model has demonstrated to the world how to achieve sustainable marine development by working on administrative and scientific fronts. This project has also made Xiamen more determined to continue this path.

In ICM, the Marine Management Office has played a very important role. Under the Chinese system, work could only be done with sound coordination. If the agencies involved went their own way, then nothing could be achieved. In this process, the Office has worked as a coordinator. It had many branches and those branches would meet regularly to discuss problems every month. In retrospect, setting up a coordinating agency has been very important for the success of Xiamen's ICM.

※ The Marine Management Office smoothes over management issues

At that time, I was also the leader of the Xiamen Environmental Protection Bureau. Working for the Bureau was intersected with many interesting stories.

In the early days, the Environmental Protection Bureau was responsible for ocean pollution monitoring, but couldn't handle it, because trash floating on the sea is difficult to deal with. Later, at the Marine Management Office, we decided

1 3

2

1. Law Enforcement Officers were documenting facilities before removing fishing rafts.
2. Drones were launched to monitor island ecology and guard the sea.
3. Law enforcement at sea

that land pollution was the responsibility of the Environmental Protection Bureau of Xiamen, whereas the Xiamen Bureau of Oceans and Fisheries took charge of sea pollution. The Marine Management Office monitors pollution, and then the actual sources of pollution are reported to its sister departments. In this way, the marine environment can be effectively maintained.

In the old days, whenever a typhoon arrived, trash would heap up on the shore, making the whole coastline very dirty. Later, it was handled by the Marine Management Office. A marine environmental sanitation department was set up by the Municipal Parks Bureau of Xiamen. They set up a team and commissioned a boat to deal with ocean trash. This work continues to this day.

Sometimes departments can be more effective when they cooperate. Take marine pollution accidents treatment for example. The Environmental Protection Bureau's research into the land sources of marine pollution could help us better grasp the overall problem. I remember that when a deputy mayor led a team to Xinglin Bay to inspect local development and construction, fishermen surrounded him on the boat and reported the death of their fish. An official immediately called me and asked the Marine Management Office for help. I took a quick look at the problem. First, the polluting factories that the fishermen reported were relatively far away from the aquaculture areas where the accident occurred. If the water was polluted by factory sewage, there would be dead fish along the route from the aquaculture area to the factory.

However, no dead fish were spotted close to the factory, therefore, the dead fish could not have been caused by factory pollution. Second, after we found out that the dead fish were all of a single species, experts from the Marine Management Office decided that this was a targeted infection rather than pollution. Through the coordination of various departments, the Marine Management Office clarified all the problems in the area, and the fishermen returned home satisfied. Think about it, if the Xiamen Municipal Bureau of Oceans and Fisheries handled the case alone, it would have approached other departments to conduct some investigations. This would be very troublesome and time-consuming. Therefore, the Marine Management Office can smooth things over when dealing with the pollution accidents by coordinating departments and division of labor.

Through the Marine Management Office, we will also coordinate with each other, making the whole Integrated Coastal Management more scientific and reasonable.

Take laws and regulations formulation for example. Administrative departments that formulate local laws and regulations in Xiamen often tend to focus on their own interests in the formulation process. With an institution like the Marine Management Office, regulations on coastal management can be formulated in a more comprehensive manner.

The Environmental Protection Bureau also has an ocean monitoring branch. Later, the Xiamen Municipal Bureau of Oceans and Fisheries and the Municipal Bureau of Parks and Woods set up ocean monitoring stations. Data gathered by different monitoring departments at different times and in different sites were different or even contradictory. Therefore, the Marine Management Office conducted coordination so that all monitoring departments could better cooperate and share data. If the data were contradictory, all departments should coordinate and figure out which data were accurate. Therefore, the Marine Management Office does play a significant role on multiple fronts.

I believe that the Office has achieved some major accomplishments in the past 20 years. The Office has helped publicize coastal management experiences of Xiamen to the world, developed Xiamen into a city planned around ports and established a science-based and reasonable functional positioning for coastal management in Xiamen.

In addition to administrative branches, this Office also has a Marine Expert Panel. Sometimes the Panel has divided opinions. However, different from administrative branches where conflicts arise due to conflicting interests, the Expert Panel has disagreements due to varied

◄ A cutter suction dredger working in the sea (photo / Wang Huoyan)

scientific and logical reasoning. For example, some experts on biological research suggested mangrove forests planted to preserve coastal zones near the Gaoqi International Airport. I expressed that mangroves were characteristic of increasing biomass. Additionally, mangroves attract many egrets and seabirds, and they would jeopardize airplane safety. Experts were persuaded, and gave up their proposal.

Experts generally consider problems from the angle of practical science, while as administrative officials, we tend to underscore the opinions of the municipal government. But when proposals submitted by the expert panel disagree with those of the government, we must take the whole picture into account.

※ Eastern and western sea areas: different locations, different requirements

Environmental protection means a lot to ICM. At that time, the Marine Management Office had three clear functions. The first was establishing scientific methods for marine development in accordance with environmental impact reports. For example, we need to be meticulous and attentive to functional orientation and potential impacts on tidal flow when it comes to land reclamation.

Efforts to implement ICM in Xiamen have been very successful, and we hope that the work can be improved in the future. Although we have already completed the preliminary international projects, we need to bear more results. Some management problems still remain. For example, in the past, some experts have resolutely opposed land reclamation. However, this view is narrow from my perspective. Therefore, I made more efforts in coordination.

When implementing ICM in the western sea areas, I argued that the area was paramount to developing marine economy. The key is to dredge the port, dig deep while still prioritizing environmental protection.

If the western sea areas are well protected, economic development is bound to ensue. But if someone proposed land reclamation in the western sea areas, I would not give way. On the contrary, I would be more open to land reclamation in the eastern sea areas. At that time, the tidal flows of this region were in good shape. At that time, I argued with our director on whether the sea area in front of the Conference and Exhibition Center should be reclaimed. The Center was originally planned to be located in the countryside, far away from its final location, which would hinder development due to inconvenient transportation. Reclamation was called for after finalizing the present site. At that time, many experts held that  land reclamation would cause pollution. They put forward two arguments. One was that the resulting tidal changes might negatively affect nearby Huangcuo beach. And many people in Jinmen were against reclamation because they thought it would cause sediments to be carried to Jinmen, affecting their sea area. These two problems were real, but I still supported this proposal. After we inspected many locations, it was the only location suitable for the Exhibition Center. But in the long run, with cross-strait relations improving, Xiamen and Jinmen can cooperate to treat sedimentation. Later, I suggested building a cofferdam before land reclamation, and the builders agreed.

In fact, in the Integrated Coastal Management of Xiamen, all work keeps pace with the times. For example, now, during coastal development, we need to consider how best to contain floodwaters when the flood comes. In the past, we only considered pollution, and never thought of flood discharge. This made us encounter many setbacks. Now, we have to learn from past lessons.

※ Looking at problems from a historical perspective

Xiamen is currently developing Tong'an Bay, and it involves a great deal of reclamation. In my opinion, in order to further develop transportation, it is necessary to conduct land reclamation. But I suggest that we should not do that without a plan. In some areas, land reclamation is easy, but afterwards, the tidal flows will be affected. It is advisable to avoid sedimentation as much as possible and make sure tides flow. Besides, tidal influx should remain unchanged after reclamation. Therefore, in this case, we also call for experts' opinions.

▲ A new and beautiful residential complex by Xinglin Bay

Land reclamation remains contentious in ICM. In the past, the Expert Panel often disagreed with the administrative group on this matter and I need to coordinate in this process. For example, in the western sea areas, I support the expert opinion that reclamation should proceed only after very careful and strict review. While in the eastern sea areas, I support the administrative group. The experts stuck to their opinions. But after all, administrative departments should consider all kinds of problems in the early stages and conduct coordination. Take the Exhibition Center mentioned above for example. Many experts opposed reclamation, but in the end, the reclamation did not exert many negative effects.

Sometimes, however, compromise should be made between experts and officials. For example, in Tong'an Bay, from a science-based perspective, the reclamation area should not be immense, but for economic development, reclamation was necessary. Another typical case is Xinglin Bay. Some different opinions still existed. Experts suggested that although the area was already silted and could be filled, but after reclamation, the rate of siltation might accelerate further. But concerning urban development at that time, land reclamation was mandatory.

In retrospect, some people may say that our previous work was insufficient and not environmental-friendly. I think that view is wrong, because in the early days, most people did not understand the concept of environmental protection, and all the plans were made for economic ends. Things should be seen from not only a science-based, but also a historical perspective. I often tell others to have this kind of perspective, and analyze both the merits and demerits of our decisions, without jumping to conclusions simply by today's standards.

Strong Science and Technology Bases for Integrated Management

Hong Huasheng

As the first female doctor of oceanographic science to return to China from overseas study, Hong Huasheng has been recognized as one of the leading oceanographers at Xiamen University. Over the past two decades, a team of scientists led by Hong Huasheng, have offered strong science and reliable technological support for the Integrated Coastal Management in Xiamen.

"All work must be forward-looking and be guided by science. " In interviews, Hong Huasheng often emphasizes this principle, which has also been her approach to years of research on the oceans. Thanks to the efforts of scientists like Hong Huasheng, Xiamen has successfully formulated the Marine Functional Zoning and has become a model both at home and abroad.

Besides her work on Integrated Coastal Management in Xiamen, Hong Huasheng has also spared no effort in raising public awareness about marine environment. Thanks to her efforts, the World Ocean Week in Xiamen, a platform for international exchange, has greatly contributed to the sustainable development of coastal zones by increasing awareness of integrated management efforts.

The ultimate goal of scientific research is to improve people's lives. Livelihood is what concerns Hong Huasheng most. "I'm a scientist, and it's my duty and responsibility to make contributions to the development of the marine economy, the protection of the marine environment, and to enhance people's livelihoods. " This is Prof. Hong's firm belief about her beloved marine science.

※ Set up a Marine Expert Panel

The Integrated Coastal Management in Xiamen reached a turning point in 1994 when rehabilitation began on Yundang Lake. How did this program begin?

In 1992, the United Nations Conference on Environment and Development passed Agenda 21, which devoted a whole chapter to the sustainable development of oceans, and mentioned Integrated Coastal Management. It was a rather new concept in China back then. An international project on the integrated management of oceans was designed to be piloted somewhere on the East Asian seas. Xiamen was chosen when this project came to China. On the recommendation and with the support of the State Oceanic Administration, the Xiamen Demonstration Project for Prevention and Management of Marine Pollution in East Asia seas was co-launched by the Global Environment Fund, the United Nations Development Program, and the International Maritime Organization, which was the starting point for Integrated Coastal Management in Xiamen.

As a transport hub in East Asia on the western coast of the Taiwan Strait, since ancient times, Xiamen has been an important port of trade connecting coastal regions in southeastern China to the outside world. Since the inception of Reform and Opening up, Xiamen has become an iconic port city with famous natural scenery, and one of five special economic zones in China. Oceans and coasts are the most important economic and biological resources and lifeline for Xiamen, and they have played a critical role in the city's economy. The fast economic growth in the 1980s and 1990s once put enormous pressure on Xiamen's coastal environment. Only by protecting oceans and coast properly can Xiamen grow sustainably. As a result, how to best use coastal marine resources and tackle environmental problems has become a headache for Xiamen's sustainable development. Luckily, Xiamen's senior government officials then were open to new ideas and hoped to change the old ways of coastal management in Xiamen.

Awareness and international projects and concepts alone are not enough. What turns theory into reality is practical application. We could not copy other countries without second thought in terms of Integrated Coastal Management.

New management methods only come about when advanced science and new technology have been put in place. At that time the Disciplines of Oceanography in Xiamen University had already become well-established, but the problem was that researchers hardly cooperated with government agencies. When the Xiamen municipal government began implementing Integrated Coastal Management, the government decided to set up a Marine Expert Panel and invited me to head this group. I didn't agree at first, because I was a scientist and knew very little about public affairs. But later I was told that my background in oceanography could strongly support the implementation of Integrated Coastal Management. I decided to get on board as I was hopeful that the program could help protect the seas near Xiamen.

The municipal government had set up several working groups of experts, including a group for civil engineering a group for science and technology. I suggested we needed only one high level expert group made up of authoritative members, which could spearhead coastal management specifically. Another suggestion was that, besides university professors, a certain proportion of members must be government and administrative experts. Based on these two suggestions, a special marine expert group was set up and, different from today's working groups, included experts with multiple-disciplinary background.

※ Develop and ImplementMarine Functional Zoning

The first thing we needed to do upon setting up the Marine Expert Panel was to figure out the proper way to delineate marine functional zones in Xiamen.

The marine environment has multiple layers, and it is inevitable to see a variety of industries develop in a one sea area, which is likely to create conflicts between the different participants as they engage in the development of shared marine resources. There were conflicts between shipping, fishing and tourism industries, and between sea reclamation and marine conservation. All these conflicts could result in unregulated and excessive development, which in turn could lead to depletion of marine resources, worsening pollution, and even the collapse of the marine ecosystem.

Marine Functional Zoning avoids blind, unregulated, and excessive development, and contributes to wise decision-making and sustainable utilization of marine resources. It is therefore a cost-effective method to tackle conflicts between different uses.

Introduction to the speaker

Hong Huasheng, Distinguished Professor of Xiamen University, Honorary Director of the State Key Laboratory of Marine Environmental Science (Xiamen University), chief scientist of the Coastal and Ocean Management Institute (COMI) of Xiamen University. In 1995, she served as head of Xiamen Marine Expert Panel and deputy director of Xiamen Coastal Zone Sustainable Development Training Center. She has participated in and led the development of Xiamen Marine Functional Zoning and the sustainable development of Integrated Coastal Management.

▼ After Xiamen marine functional zoning, Southeast International Shipping Center located its headquarter in Xiamen Haishu, which is the fourth international shipping center in China after Shanghai, Tianjin and Dalian. (photo / Wang Huoyan)

Marine Functional Zoning is defined as a decision-making process to formulate effective mechanisms and solutions to optimize the distribution of resources. Its main purpose is to regulate human activities, so as to minimize disturbances to the marine environment and while satisfying people's demand for consumption and production demand through a functioning marine ecology.

However, the marine economy is a large-scale interdisciplinary system that involves multiple regions and industries, which since the early days has made the management and coordination of marine areas more complex than land. In addition, all the government agencies involved in coastal management in China acted according to the powers granted by relevant laws and regulations of their respective branches. Their respective needs were their only concern when they formulated and implemented marine development plans. The lack of coordination and testing mechanisms inevitably led to chaotic development and conflicts between different uses in the shared area, which resulted in the waste of financial and labor resources. The underlying cause of these problems was flawed legislation on coastal management. Therefore, it was also very important to establish sound laws and regulations related to Marine Functional Zoning, equivalent to those for land management.

The Marine Expert Panel and sea-related agencies held ten meetings to formulate a detailed plan for marine functional zoning. At the last meeting, the then Xiamen mayor was present, and after we reported the plan to him, he immediately informed us of the approval of the plan by the Xiamen government. In 1997, the plan was passed by the municipal People's Congress, which gave it legal authority. Looking back on it 20 years later, we are glad to see our concerns back then were well addressed in the plan.

※ West Sea Areas: Give place to port industry

Marine zoning basically delineates the proper usage of marine resources. Xiamen was the first city in China to begin marine functional zoning. In the early stages, development of the seas was chaotic. For example, both fishing and shipping were conducted in one area, creating many conflicts. It was urgent to begin Integrated Coastal Management to improve the functioning of sea areas.

In order to formulate plans for Marine Functional Zoning, together with some experts, I went on a research tour of the current sea usage in Xiamen, and we paid special attention to the coastlines and ocean currents, before we conducted further study on issues related to the usage of sea resources, such as the impacts of vast sea reclamation. The research team carried on with the issues concerning the future development of Xiamen, and how seas

▲ Crowded cages under Haicang Bridge in the old days

of Xiamen should be used. The Marine Expert Panel established the framework, which defined the use high priority, maximum benefit for multi-resource uses, and restricted functions of each sea area.

We started work in the western sea areas first. The traditional industry in the western sea areas is aquaculture. At that time, there was only a small Port at Dongdu, and Haicang Port had not yet been built. On flights to Xiamen, foreign guests might be confused to see white patches dotting the western sea areas. In fact, those were patches of foam farmers laid in the sea for fishing. At that time, the port industry in Xiamen was emerging as an economic pillar. However, transportation channels in the entire Dongdu area were clogged by aquaculture. When entering Dongdu Port, ships were often ensnared by fishing nets.

Consequently, it was important to answer the defining question: what was the major purpose of the western sea areas? Despite the large number of people engaged in aquaculture, we suggested the port industry should be prioritized in the western sea areas, since the Xiamen government had put forward the vision of "building an international port city with beautiful scenery". Accordingly, only by giving ports top priority over any other activities in the western sea areas could the port industry grow in Xiamen. Thanks to this vision, Xiamen has grown into an international shipping hub.

Ports serve as the dominant function of the west sea areas, with tourism serving as a compatible function. Dayu Island was already under development at that time. The city built mangrove forests there and the island became home to egrets, and the scenery was gorgeous. Because a waste water treatment plant was located in the western sea areas before our planning began, discharging of waste water was also included as a compatible function in the early version of Marine Functional Zoning. Later, this function was gradually phased out.

Aquaculture was the industry that required the most urgent restriction in the western sea areas. We worked hard to remove it. At that time, sea-related government agencies in Xiamen reached a consensus that the port industry in the western sea areas should be developed, but disagreed on the necessity of phasing out aquaculture. Experts from the Marine Expert Panel tried their best to persuade the leaders. We raised two main arguments. The first was that aquaculture was likely to impede the construction of ports and navigation in the western sea areas. The second was that aquaculture would lead to the deterioration of water quality, leading to the industry's own destruction. Aquaculture required Type-2 water, while shipping, which only required Type-4 water, was quickly growing. When the quality

※ Oceanic research still has a long way to go

The ocean is complex and ever-changing.

With talents from different backgrounds working together, ocean can be best managed and utilized, and developed. Work done in Xiamen has a solid basis and will develop at a faster pace with the establishment of the South Ocean Research Center and strong economic support.

In my opinion, besides fulfilling due responsibilities, the municipal government and academic institutions should foster communication with institutions across the world. Since the 1980s, Xiamen University, the pillar of ocean research in Xiamen, has worked continuously with

Taiwan on researching the ecosystem of the Taiwan Strait. New problems keep emerging during development. Since the late 1990s, global warming has had severe impacts on the ocean, and we need to work together to address it. Another example is carbon cycling—how do marine organisms make use of carbon and store it in the ocean? This is a long-term research project requiring efforts of generations.

Taking my most familiar marine planktons as an example. At the early stages, our research was tangible, but it has completely changed now. We are at a higher research stage and have to rely on modern technologies like genetic technology. Microorganisms play a major role in the ocean ecosystem and their influence exceeds that of large organisms in the sea. Research based on the ecological variations of microorganisms is a key approach to ocean research, and will take ecological research on marine planktons to an even higher level.

Despite the progress made in the theoretical and technological fronts of ocean research, it remains a broad and complex topic. Humans still has many uncharted territories in the ocean. Xiamen Bay, a place that many people find disgusting due to sea waste, has made people wonder why it remains foul after treatment. But if we look at it from a global perspective, a massive plastic island has formed in the depths of the Pacific Ocean, a situation far worse than Xiamen Bay. Yet we still have no idea how the waste floated all the way to form this island. I think our research should be expanded from coastal research to better encompass the open ocean and the undersea areas, so that we can gain an integrated knowledge of the rules of nature and act accordingly.

The ocean—once a "shopping basket" to a "test field" for people here.

From my point of view, Xiamen is doing a great job in developing its marine economy. Xiamen has based its

▼ Sunset on the coast (photo / Lu Qihui)

development on advanced and new technologies to make up for its relative lack of resources, especially in terms of promoting the medicinal use of marine organisms. With the hi-tech park, Xiamen will incorporate other technologies in the future to promote sustainable development in every respect.

While it appears that fishing in the open ocean is not cost-effective at the moment, Xiamen should act from a long-term perspective, as it is a potential solution to problems concerning people's livelihood. Even though Xiamen's traditional fishing industry is transforming into an urban fishing industry geared to sightseeing and leisure, this is far from enough for local fishermen. The future development of fishing industry here needs to be further discussed and explored.

Apart from this, we should work to raise awareness of the issues related to the ocean. Several departments and institutions have made such attempts, but we need a more integrated approach. Another important work is unified use and dispatching of ships. Progress has been made and different departments and institutions should work to communicate better and alter their attitudes towards the administration of coastal areas.

All in all, relevant work should be driven by science and technology, and Xiamen should put more efforts in this area. The building of the South Ocean Research Center has proposed that science and technology propels productivity and more support should be placed into its development. However, we should never give up on investing in fundamental research. After all, fundamental research provides breeding ground for nurturing ocean research.

▼ Dayu Island is an important breeding ground for Xiamen Egrets, and a provincial "Dayu Island Natural Reserve of Egrets in Xiamen" has been established there. (photo / Yao Fan)

Coordination Between Marine and Land Administration to Propel Ecological Rehabilitation

Yu Xingguang

Against a vast blue sea, the waves tap the white beach rhythmically and the warm breeze carries with it the fragrance of salty sea water. Since 1980, Yu Xingguang has worked at the Third Institute of Oceanography of the State Oceanic Administration, and has led a life surrounded by the sea for 36 years. He passes by the sea on his way to work and can watch it out the window of his office. The sea is just at his fingertip. It has given him boundless space for imagination and many fond memories.

"The sea near the Conference and Exhibition Center was just splendid! I was much impressed", Yu Xingguang recalled reminiscently. It seemed as if he could have the immense stretch of beach and the sound of waves all to himself and he could spend time with them whenever he liked.This was a gift bestowed by destiny. He said, "Xiament could not even count itself as a city for international tourism if tourists cannot get near the sea. "

In 1956, Yu Xingguang was born in Hetian town in Fujian along the Longyanting river, less than 300 kilometers away from the picturesque coastal city of Xiamen. Out of his affection for the sea, he left his hometown in the mountains and marched towards the seaside. In 1980, he graduated from Sichuan University and began to work in oceanic science. The sea, once a far-away dream, became readily available, which was quite a wonderful experience for him.

At the sight of trash floating at the sea after a downpour, he feels distressed and the urgency to accelerate pollution treatment in the Jiulong River basin. He emphasizes the importance of ecological protection in government policy making. One of the letters he once wrote helped Wuyuan Bay wetland to be conserved. He felt a sense of happiness when recalling the catalytic role he played. As an outspoken marine expert and a deputy of the Municipal People's Congress for the past 20 years, he has raised many important bills, covering integrated management of the western sea areas, treatment and management of Yundang Lake, integrated management of the eastern coast, beach and wetland conservation, national ocean park construction, soil erosion prevention and marine environmental protection, pollution treatment in the Jiulong River basin, ocean cultural promotion, ecological conservation at the Xiamen International Garden & Flower Expo, and Causeway renovation and ecological rehabilitation. He has been dedicated to ecological rehabilitation work in Xiamen. He once remarked, "Our knowledge of the sea is quite limited. It takes generations of efforts by scientists to explore, study and discover the secrets of sea. "

With his deep affection for green mountains and blue sea, he has been advocating the concept of marine and land coordination in marine environmental treatment and ecological conservation. He said that a coastal city should keep up with the times and integrate the sea, mountains and land management. From his view, the entire construction process of Xiamen from an island city to a coastal city is synonymous with the process of ecological urban construction.

From better understanding to concrete action, coastal management in Xiamen has undergone an extraordinary transformation. According to Yu Xingguang, "we were creating international precedent and Xiamen should harbor a sense of pride." Throughout these years, Yu Xingguang has promoted the Xiamen experience in numerous places and advocated for ecology, wealth and tourism related to the sea.

※ When the environment is clean, egrets will return

To improve the marine environment, Xiamen has launched numerous coastal management and protection programs. From pollution treatment to ecological protection, to the construction of an "ecological civilization", the Xiamen experience proves valuable for its continuously evolving guiding principle.

Introduction to the speaker

Yu Xingguang, former director of the Third Institute of Oceanography of the State Oceanic Administration, Ph.D., Researcher, Ph.D. Supervisor; State Council Special Allowance Specialist, Member of Xiamen City Technical Elite; Director of Marine Ecology Committee of China Ecological Society, and Director of Environmental Ecology Committee of China Ocean Engineering Consulting Association. Since he joined the Third Institute of Oceanography of the State Oceanic Administration in 1980, he has devoted himself to marine environmental protection, actively promoted marine ecological civilization, and presided over the " The Jiulong River Basin-Estratery Ecological Security Assessment and Regulation Technology Research", among other major national public welfare projects.

▼ Egrets are flying around in Xiamen, which is known as the Island of Egrets, attracting lots of tourists. (photo / Wang Huoyan)

The sea in Xiamen only covers an area of 390 square kilometers. So how can its experience be of use to the entire country? Frankly speaking, the volume of marine resources and diversity of coastlines in any province can easily overshadow those of Xiamen. However, Xiamen boasts a refined and ecologically sound coastal management model. Behind every coastline, every bay and every sea area is a customized solution of ecological construction and environmental protection.

Yundang Lake experienced a rather excruciating restoration process and still has some remaining problems. One day after work, I ran into a farmer buying vegetables. He said the sea was pretty magnificent now. But Yandang Lake still had some problems, and he had just bought a house there.

I consoled him and said that the housing price there was pretty high. With that, he waved his hand and responded that it could be higher but for the poor water quality there.

Right now when summer comes along, a foul odor will emerge from Yundang Lake. Xiamen has taken great pains in water treatment, yet has not addressed this problem. What goes wrong? Lake pollution treatment is one of the most intractable problems in environmental protection. A polluted lake is just like a patient suffering from both hyperlipemia and cerebral thrombosis, in need of constant operations. A heat stent may resolve some dynamic issues and expand blood vessels. Less intake of nutrition may yield some results. However, none can eliminate the root cause in water flows. Actually, it is also attributable to historical reasons. You could not expect people at that time to have this awareness of marine and land interactions and regional ecology, or the concept of sustainable development. This is a concept that has taken several decades to emerge.

Each river will find its way into the sea. Water in neighboring areas all flow into Yundang Lake. Right now, large-scale wastewater treatment has ameliorated the situation. The treatment process has spanned several decades, and there was a period where no single egret came back within 7 to 8 years. Birds are like human beings. They also like a clean environment. When the environment is bad, they will leave. And when the situation gets better, they will return.

I once raised a bill during the People's Congress, hoping to extend our administrative legislation's jurisdiction to Songbo Lake which was situated upstream of Yundang Lake. At first, this lake was not incorporated into the administrative region of Yundang Lake. Then during the treatment process, problems also sprang up in Songbo Lake and people living around this lake voiced their complaints. Then the pollution and stench treatment efforts of Songbo Lake were launched. These experiences have taught us valuable lessons. Actually, these lessons have informed decision makers and the general public of the state of the marine environment.

The treatment in Yundang Lake lies in treating stench and silt which affect the cityscape, regulating the regional ecology, addressing citizens' complaints, and making the coastline more accessible. The Xiamen municipal government has conducted follow-up work for over 3 decades, and the project is recognized as a demonstration project by the UN's relevant bodies, a hard-won result. My teacher, Professor Lu Changyi from Xiamen University, organized a mangrove planting activity in Mid-lake Island, which has proven a success. I also engaged in volunteer work to plant trees.

※ Opening up Causeways brings cross-regional development

I have just talked about the issue of water flow. How can we ensure the smooth flow of oceans, rivers and lakes? As for Xiamen, it needs to open up the causeway. This large operation is a good example and is a testament to the hard work of Xiamen people. We did not deny that the causeways made tremendous contributions to Xiamen"s development in history. But, in the past, economic conditions and technical expertise were limited and we could not expect builders at that time to work out mathematic models.

This project was under the charge of the then Vice Mayor Pan Shijian and facilitated by the Third Institute of Oceanography of the State Oceanic Administration and Xiamen University. The Nanjing Hydraulic Research Institute conducted physical modeling experiments to decide the width of the opening? Should it be 800 meters, one kilometer, or wider? After full corroboration, the opening was finalized at around 800 meters. The opening was made to comply with natural laws to restore the natural dynamics of the ocean. This case of ecological rehabilitation serves as a wake-up call for the world, especially developing countries. It is inadvisable to enclose the sea at random.

In 2002, Xiamen proposed building itself into a coastal "ecological city" and basically realized this objective in 2007. So how does a coastal ecological city come into being? First and foremost, a well-thought long-term plan is needed. Some places might just build a causeway to withstand storm and coastline erosion. When a typhoon strikes, the coastal shelter-belt is blown down and the coastline and beaches get eroded. Then what? Another causeway is built. The coastline might be protected, but the beaches are ravaged, which means the loss of major coastal tourism resources. All this boils down to insufficient understanding of the sea. Xiamen did not go down this road. Its eastern coastal management is a much larger endeavor. This integrated management project covers 114.28 square kilometers of land and about 91 square kilometers of coastal area. The coastline involved amounts to about 58.5 kilometers. The project intends to clean up 220 million cubic meters of silt for rational use.

If the resolution had not been strong enough, the region would have remained in the same sad state. Xiang'an, Tong'an, and Jimei might not have been incorporated into Xiamen. It would have had to make do with half-measures like dredging. But this is not what happened. The project was instrumental in expanding urban areas, alleviating transportation, population and land resource strains, and even promoting a higher level of ecological tourism.

But for the broad structure, how could we have accelerated the development in Dadeng and Xiaodeng Island areas, not to mention larger areas like Xiang'an, Jimei, and Tong'an? In a sense, it was strategic thinking and action—making full use of theories of ecological management, drawing experience from other major coastal management efforts, and even monitoring dredging.

Why was the monitoring necessary? During construction, there was a gap between the original plan and our actual work. Some measures were not up to standard and some units failed to comply with relevant regulations. Some workers might dump the dredged mud halfway and find a random place to dig sand when the supervisor was not looking. This inordinate sand dredging exerted huge pressure on marine environment. How were we to ensure that there would not be new pollution and environmental problems in the treatment process? Xiamen Municipal Bureau of Oceans and Fisheries and the Construction Headquarters devoted a lot of efforts, for example scanning the area of the sea both before and after the dredging so as to count the exact amount dredged. In 2007, during the first Session of the 13th Municipal People's Congress, 10 other deputies and I proposed a bill *Reinforcing Supervision to Realize Integrated Management and Ecological Rehabilitation Objectives in the Eastern Sea Areas*. This bill was listed into the agenda of its standing committee. The Xiamen Municipal People's Congress Urban Construction Environment and Resources Committee has conducted a two-year supervisory effort.

On the surface, the project aims to turn Xiamen into a coastal ecological city. In reality, it intends to open Xiamen up to a large region and assuage local development pressure to some degree. At the same time, it broadens the interactions within the city and between the city and other areas. Therefore, this kind of broad transportation network has laid the groundwork for the next twenty or so years' development. Different strategies lead to different results. On the surface, Xiamen remains a coastal city, despite the project, but

▲ Kids by the sea (photo / Xu Jinyue)

in reality, the changed philosophy bring huge strategic benefits. Xiamen attaches vital significance to ecological conservation and has conducted such projects as coastline management, afforestation of mangroves, and returning land to the sea. These foresighted measures carry ecological significance.

※ The disruption of ocean currents led to the demolition of the Sea View Building

There was a period of time where, to my chagrin, because of construction, many coastal beaches were destroyed. I believe that for many, the following story will bring about feelings of nostalgia.

The Chinese name "Wang Hai", which means "gazing into the sea", sounds quite poetic, and unleashes the romantic imagination of tourists. The Sea View Building was originally situated under the hilltop of Third Institute of Oceanography of the State

Oceanic Administration and was built on the beach. The designers of the Sea View Building were well-meaning. It was meant for both catering and sightseeing. However, its location had a large impact on the local environment and resulted in the redistribution of sand, which altered the form of the beach. Previously, it was a half-moon-shaped bay. Gradually, the big half-circle vanished and new problems emerged. The coastal management department had to erect a sign which said "Beware of deep currents when swimming!"

Subsequently, the municipal government became determined to dismantle the Sea View Building due to suggestions solicited from experts and citizens, who believed that the area should be revitalized through the demolition of Sea View Building and subsequent construction of a bridge, removal of waste water and creation of new scenery. They hoped to restore the beach according to Xiamen's reputation of having beaches fit for swimming. The Sea View Building

was a must-see for many tourists coming to Xiamen. In an instant, it was gone, causing a palpable sensation among the public. Xiamen took this matter seriously and activated a sand rehabilitation system. It adopted different measures according to the specifics of each situation and this sustainable action was quite worthy of emulation and praise. One year, under the aegis of Xiamen Municipal Bureau of Oceans and Fisheries, I led an expert panel to conduct thematic field trips on beach protection. I remember it was a day before Chinese New Year's Eve and municipal government leaders took time from their busy schedules to hear the research report, which we found very touching.

In the final analysis, the results of beach management are determined by the advances in planning and theory. This reminds me of the fact that the construction of the national-level Ocean Park stemmed from a casual conversation.

▲ Open the causeway is aimed to restore the natural dynamics of seas, following the law of Nature.

One time, I went to the airport to pick up Mr. Zhouzheng, Political commissar of the China Marine Surveillance Fleet. Zhouzheng joked that Xiamen should erect a commemorative arch or sign near the airport to remind tourists to visit the Ocean Park. Actually, this incident became etched in my memory. The idea to build an ocean park of Xiamen's own had been brewing in the relevant departments for quite some time. Finally, an opportunity came along.

During the first half of 2010, I went to Guangzhou to attend a meeting held by the State Oceanic Administration. I learned that the SOA would issue *Management Regulations for the National-level Ocean Park Application*. I got a document on soliciting opinions as well. Once the meeting was concluded, I set out for the North Pole to conduct an expedition. And with the support of the SOA, I handed over the work to 2 colleagues from the Third Institute of Oceanography of the State Oceanic Administration to compile the plan for the National-level Ocean Park, and it needed to be completed fast. My idea was to incorporate the area stretching from Huandao Road to Xiamen's swimming beaches into a large "ocean park loop". It was positioned to make it into the first batch of National-level ocean parks. It was indisputably worthy of this title for the abundant coastal resources like Hulishan Artillery, its long coasts and beautiful beaches, and the Wuyuan Bay tourist area. Shangyu Islet is also a scenic attraction, which was situated on the coast of the Exhibition Center, 4 kilometers away from Kinmen Islands in Taiwan. The main section of the Huandao Road boasts an advantageous location in the re, as it is at a prime location for gazing out into the Taiwan Strait.

It came as a surprise that when I reviewed the approvals of the first National-level ocean parks after I came back from North Pole expedition, I did not see the name of Xiamen. Xiamen had not submitted the application. I was rather anxious upon receiving this news and learned that the case did not pass the municipal government meeting. I participated in the meeting held later as the representative responsible for plan compilation. The motion was barely denied by Mr. Liu Cigui, the mayor at that time, reasoning that not the single lane, but the whole Xiamen should be considered a national ocean park. Considering the environment of Xiamen, this notion was understandable. But at that time, constructing such a large ocean park was not realistic. It did not correspond with the standard of national ocean park construction and would be difficult to incorporate into the grand scheme of a national ocean park.

The Huandao Road region has what it takes to become a national-level ocean park. The reasons are numerous. Each year, Xiamen International Marathon is held there. Being close to Jinmen, it still bears the scars of war and thus has historical significance. It connects Wuyuan Bay, the International Exhibition Center, Artillery, and the "bathing" beaches. I felt really grieved that Xiamen, as a renowned coastal city, could not be crowned with the national title. Later, the municipal government agreed to file an application. Mr. Liu Xigui remarked that it would be a shame if Xiamen only made the list in the second round, showing a determination to make it to the first batch. The officials from the Ocean & Fisheries Bureau and I went directly to the SOA and approached the department in charge. We made an impassioned effort to demonstrate our case and our hard work paid off. Finally, Xiamen made the list into the first batch.

Shortly after Mayor Liu Cigui was transferred to take office as the Administrator of the SOA, he approved Xiamen's application as a national ocean park. On May 19, 2011, Xiamen National Ocean Park was inaugurated and was listed as one of the 6 national ocean parks in the first batch. Elevating its status to national ocean park was a pivotal step in the efforts of resource protection, coastline rehabilitation and bonding together the general public. If a city can have more marine elements, it will pay dividends. I live in Qianpu and often walk to work, and walking to work in a "national ocean park", is quite enjoyable.

※ The Wuyuan Bay and Xiamen Yuanboyuan Expo Garden are impressive engineering projects

The treatment of the Wuyuan Bay taught us an important concept, which is that protecting the natural landscape brings great ecological dividends. Treatment and rehabilitation are not isolated efforts. Key ecological marshes should be conserved as much as possible because the future social significance and functions will be tremendous. During the treatment process, one freshwater marsh was slated to be turned into real

New and old Wuyuan Bay

estate property. Experts from the Third Institute of Oceanography of the State Oceanic Administration were assigned to conduct an ecological survey. So I wrote a letter to Mr. Pan Shijian, Deputy Mayor, and placed two photos inside, one depicting a buffalo lying in a small pond with two cattle egrets standing on its back, and another depicting a cow eating grass surrounded by egrets and aquatic plants. To the photos, I added that one day the two photos might become the reminder of harmony between human and nature in Xiamen. I hoped to elicit field trips and construction plans. When Mr. Pan Shijian received the letter, he passed it on to the Xiamen Urban Planning Bureau. It accorded vital significance to the matter and ensured the protection of the dozens of hectares of

freshwater marshes. When I stood there, I felt greatly delighted and proud. Subsequently, when the Conference and Exhibition Center displayed pictures of the best residential areas in Xiamen, the photos were of similar types.

Rehabilitation of both the inner and outer portions of the Wuyuan Bay reflects advanced guiding philosophy. The external bay will be made into a tourist attraction and yacht base. The inner bay will be made into an ecological marsh and form a new scenic spot. Marine treatment, rehabilitation, and protection need technical support and cannot be achieved unilaterally in a "Captain's call". Technical support reminds decision-makers to take note of the laws of nature, or experiences and lessons from home and abroad. Xiamen coastal management is backed by science.

The Xiamen Yuanboyuan Expo Garden is also an impressive engineering project. Mr. Wang Chunsheng, Director of Xiamen Municipal Bureau of Oceans and Fisheries, served as deputy commander-in-chief of the project. Once he came to the Third Institute of Oceanography of the State Oceanic Administration to solicit suggestions on development and protection, and we talked for over one hour. I proposed that the ecological elements, such as natural river systems, reed wetlands, lotus wetlands, tracts of farmland, hills, forests, and avian habitats should be preserved to upgrade the ecological quality of Xiamen's

◀ Garden Park of Xiamen, a gorgeous garden (photo / Wang Huoyan)

Yuanboyuan Expo Garden.

Mr. Wang Chunsheng was aware of these issues and assured me that those ecological elements would be preserved. This decision meant an overhaul of the whole plan, but he was determined to convince related departments. I also organized and collected materials for corroboration. After that, the department in charge conducted an ecological survey and I led the work to subscribe the bill on salvaging the ecological resources of the Xiamen Yuanboyuan Expo Garden. The bill caught the attention of Municipal People's Congress and government as well as deputies in Xinglin district. The Xiamen Municipal Environmental Protection Bureau also submitted documents to the municipal government to show their supports.

During the construction of the Xiamen Yuanboyuan Expo, the original ecological landscape was maintained and the area was developed with the help of environmentalists and ecologists. Their prior surveys helped inspired new ideas, which were instrumental to better understanding the interactions between development and ecological protection and making development more ecologically friendly. In this regard, Xiamen did quite a good job.

※ Coastal management still has a long way to go

I often advocate turning the sea into a sea of ecology, a sea of wealth and a sea of tourism. Many people come to Xiamen specifically for educational purposes. Though not endowed with the best marine resources in China, Xiamen has succeeded in winning worldwide recognition. This does not come easily.

The sea is beautiful and fragile at the same time. When a typhoon or thunderstorm strikes, it is exposed to pollutants from the Jiulong River. Some waste will flow on the water and overnight the sea's beauty will be buried in its depths. Invisible influences are significant as well, like the red tide. When you see a red sea or a sea of other colors, you may find it very beautiful. However, invisible risks are taking shape and their formation may be like a clogged blood vessel, with threatening the health of the marine ecosystem and food chain. If a red tide breaks out, ecological risks may crop up. The municipal government took this problem very seriously. Xiamen Municipal Bureau of Oceans and Fisheries set up the emergency response system and the Third Institute of Oceanography of the SOA led the work. Mr. Nichao, the former Deputy Mayor responsible for ocean administration, took time almost every day for several weeks to conduct research with experts. I also engaged in the discussions.

Marine rehabilitation has a daunting road ahead. River basin pollution is a grim challenge for Xiamen. The inner city is developing and pollution is mounting. At the same time, the external and surrounding city is flourishing and pollution there is on the rise as well. To protect the sea, the Xiamen municipal government is promoting regional cooperation. The Standing Committee of the Municipal People's Congress took the initiative and organized deputies to conduct field trips to Zhangzhou and Longyan, checking on husbundry pollution of the entire river basin as well as rural pollution and waste treatment. The three cities made joint efforts and passed the Xiamen Consensus, Zhangzhou Consensus, and Longyan Consensus.

Why should we talk about this? Xiamen is situated at the estuary of the Jiulong River, with its bountiful biological diversity. However, when pollutants flow into the sea with the river, they accumulate at the estuary. Considering the size of the sea around Xiamen, If it receives over 70% of the pollutants, it will be overburdened, in need of greater marine treatment. According to my knowledge, Xiamen has tired its best to collect and treat the pollutants. There are sewage factories in Haicang, Xinglin, Tong'an, and Xiang'an. At least 85% of urban sewage has been collected and treated, which is among the best in the whole country. Small river basins within this area are scaling up treatment efforts as well. But the pollution in the Jiulong River basin is difficult to address. It is hard to enhance the quality of the marine environment in Xiamen by a large margin. From the perspective of urban water supply and ecological safety at the estuary, I proposed a bill on this matter, which was underscored by the Municipal People's Congress, who then did vigorous work to address the problem. However, there is a long road ahead, and the national government has explicitly clarified the approach Xiamen should take to establish a marine "ecological civilization", which including the treatment of the Jiulong River basin.

Taking the Chinese white dolphin as an example, the difficulty of their artificial breeding makes their protection more difficult than giant. In this regard, Xiamen has taken many steps. Construction crews are required to ward off dolphins from explosion and collision areas, and ships to control their speed. Publicity has been worked on, and some key ecological areas rehabilitated. Meanwhile, and floating waste disposed of. The UNEP has learned of Xiamen's experience in Chinese white dolphin conservation. Scientists from the Third Institute of the SOA also work to establish a network publicizing the Xiamen's experience in conservation.

Xiamen is an exemplary case in treating all kinds of pollution in estuaries and rivers, and rehabilitating major ecological functions. The key of its experience is coordinating land and coastal management to promote pollution treatment, especially in the case of improving marine environment. Personnel from Hebei, Guangdong, Guangxi, and Hainan have come to Xiamen for field trips. Taking Hebei for example. The priority of their trip was to learn from our experience in coastline protection and management, because beaches of Beidaihe River are crucial ecological coastlines with political significance and their beaches suffer from the same degradation as ours. Now most portions have been rehabilitated. They also focus on river basin problems. To monitor pollution of the Beidaihe River and the adjacent 12 rivers, they also set up online monitoring stations and systems.

※ Xiamen's ecological culture should be enhanced

There are two important ecological landscapes in Xiamen: the coastline, and the mountains. There was a time in Xiamen, when the mountains were covered in bald patches. All mountains looked treeless. City construction and road paving left mess everywhere. I think it's a pity. The decision makers need to think about "what impression do you leave behind when the mountain is developed?"

Most people think that mountains have little connection to the sea. Actually, mountains provide protection for the marine environment in two ways. First, it is the "lung" or "kidney" of the city, regulating the air flow and serving as a key tourist attraction. Secondly, healthy mountain ecosystems prevent earth from eroding into the ocean during thunderstorms, a vital ecological function. Xiamen once went through a period of vigorous mountain excavation, and suffered a great deal from it. That's why we introduce a bill to prevent further excavation and the resulting soil erosion and deterioration of the bay environment.

As demonstrated, Xiamen features integration between mountains and coasts. Therefore,

▲ One section of local regulations, on protection of the shoreline of Haicang Bay, forbids reclamation of land in the sea and imposes penalty on damage to the shoreline. The picture shows the waterfront of Haicang Bay Park. (photo / Zhou Zanjia)

in the management of coastal cities, especially island cities, sufficient attention should be directed at the overall planning and management of coastal, mountainous and land areas. In this way, the beauty of the city will be highlighted and its cultural influence can be enhanced.

One regret is that Xiamen does not boast a cultural program with tremendous influence like Impression Lijaing, Impression Liu Sanjie or Impression West Lake. As a matter of fact, the story of the Chinese white dolphin offers a case for artistic invention, as right now, the Chinese white dolphin is still unknown to many people across the globe. In addition, Haicang Bay has a coastline sprinkled with beaches and groves, which bears large ecological significance. We could draw the attention of the general public and tourists to its beauty through education exhibitions. It can also demonstrate the commitment of Xiamen municipal government too protecting marine resources. I cherish this coastline very much and the expert panel has raised many suggestions for its protections.

The Xiamen special administrative region has legislative authority. I drafted a bill on coastline protection, hoping the People's Congress could issue management regulations, but it did not pass. Luckily, the Legal Affairs Committee of the municipal People's Congress and Xiamen Municipal Bureau of Oceans and Fisheries took the matter seriously and promoted the promulgation of several protection regulations. Among them, one paragraph concerns protection of Haicang Bay's coastline, which includes restrictions on land reclamation and the destruction of the coastline. Xiamen has walked a long way in resorting to legislation to protect the coastline thanks to the persistent petition of experts setting an example for law-based coastline management. Marine resource managements will face mounting difficulties without legal protection. From my perspective, Xiamen will need to continuously reinforce legal and ecological coastal management mechanisms.

Although around China there are all kinds of laws and regulations, regulations in Xiamen are more specific. The original pattern for *National Island Planning* sprang from Xiamen. Legal frameworks are essential in enhancing awareness among government, the general public, related builders and decision makers while regulating people's behavior.

Additionally, marine ecological compensatory legislation originated from Xiamen. Xiamen is a pioneer in the nation in this regard. At the early stage, it earmarked 20 million RMB to the upper stream region and has now raised this figure. The research report I organized and engaged in was compiled into the work, *Experience and Practice of Coastal Management in Xiamen*. For citizens in the upper stream region, efforts to protect the Jiulong River have compromised their own development. Actually, development of downstream regions to some degree benefits from the upper stream region. When citizens of the upper stream region make sacrifices for the greater good, the central financial fund should disburse a certain amount of money to the upper stream regions. In the future, to conserve marine resources, we need to integrate the efforts of both inland and coastal areas.

In summary, coastal management is a process of constantly identifying and tackling problems. Xiamen is marching ahead and strives for the best.

Multi-pronged Methods to Restore Marine Ecology

Huang chaoqun

From publicity and education efforts to the support of sea-use, the management of sea areas, and the development and utilization of sea resources, Huang Chaoqun, director of the Xiamen Oceans and Fisheries Bureau, witnessed and participated in the whole process of the Integrated Coastal Management (ICM) in Xiamen. He summarizes the Xiamen model in 20 Chinese characters, translated as: "legislation first, centralized coordination, scientific and technological support, integrated law enforcement, and public participation".

Over the past 20 years, as one of the pioneers of the Oceans and Fisheries Bureau, Huang Chaoqun has been involved in most major projects like the withdrawal of aquaculture in eastern sea areas, the integrated regulation of western and eastern waters, as well as a series of actions such as causeway opening, bay governance, and beach restoration. He oversaw a dramatic transformation of the Xiamen sea area and the continuous improvement of people's awareness of marine environment and conservation. The Chinese White Dolphin, a spirit of the sea and a symbol of the city, is now protected by the general public in Xiamen, which has created a new stage of harmony between humans and dolphins.

Pick a random area or island on the maps of seas around Xiamen, and Huang Chaoqun always has something to tell, including the general situation as well as the management process of the bay area and the island. He told us frankly that Xiamen has stepped up efforts in science, management and innovation. It has focused on allocating marine resources according to the major functions of sea areas to promote the sustainable development of marine economy. It was among the first cities to shift the focus from marine space management to ecological management. Large scale integrated management initiatives of the sea area have since been introduced, including optimizing the marine ecological structure, expanding marine area, protecting and revitalizing marine resources. By doing so, Xiamen succeeds in sustaining ecology and securing better marine functions and resources.

※ The Xiamen ICM model is one of the three major models in the globe

In the early 1990s, I took part in Xiamen's coastal management, and was in charge of publicity and education. At that time, I was responsible for raising people's awareness by educating the public of the necessity of protecting the sea, by giving out brochures and organizing counseling meetings to involve more citizens.

Xiamen, located at the mouth of the Jiulong River with 390 square kilometers of sea area and 226 kilometers of coastline, is surrounded by water. Among all the coastal prefecture-level cities, its sea area is the smallest, but its development and utilization intensity is the highest.

For a long time, Xiamen has ranked among the best in China and even in the whole world in terms of coastal management. In 1994, the introduction of ICM made Xiamen the very first in China to implement such a program. In the process of carrying out ICM, Xiamen found a method that suits its own condition and is widely recognized at home and abroad. Many of its practices and methods prove valuable experience for the implementation of ICM around the world.

In the early days, I was responsible for publicity. At that time, people from all over the country and even from other countries and regions came and learned from our experience. I was responsible for working with media that reported our ICM achievements. Every time when referring to the Xiamen ICM, we had to mention: "legislation first, centralized coordination, scientific and technological support, integrated law enforcement, and public participation".

A journalist from People's Daily and I together worked out this twenty-character summary. In 1998, the journalist came to Xiamen for an interview and asked about the Integrated Coastal

Introduction to the speaker

Huang Chaoqun, Director of Marine Division, Ocean and Fisheries Bureau of Xiamen, has been responsible for public marine education since 1991, and has been fully involved in Xiamen Integrated Coastal Management (ICM), including major projects such as withdrawal of aquaculture in the eastern sea areas, the comprehensive rehabilitation of western sea areas, and comprehensive renovation of the area surrounding the eastern seas, and a range of special actions such as opening of causeways, bay area governance, and beach restoration. He is a pioneer in the work of Xiamen Marine Fishery Bureau. Since 2015, he has been mainly responsible for advancing the "Majian Bay Basin Development and Utilization" project.

▲ There are 3 "S" to describe seaside tourism: Sea, Sunshine, Sandbeach. We have all three factors here. Come and have fun flying kites. (Photo / Wang Huoyan)

Management model during the "China's Centennial Environmental Protection Campaign". We reflected on our past experience and summed it up for him. After years of exploration, we decided on "legislation first, centralized coordination, scientific and technological support, integrated law enforcement, and public participation", a new approach to protect marine ecosystem while pursuing economic development. It is worth mentioning that it is one of the three major models in the globe.

The traditional coastal management in China is fragmented, which tends to suffer from lack of coordination and an overlapping or vacuum of responsibilities. As a coastal port city, Xiamen has 13 special coastal management departments at the central, provincial and municipal level, as well as a multitude of indirectly related management departments. Their competition for authority over resources and space is very fierce and often leads to waste or even the destruction of marine resources. To address this problem, the Xiamen municipal government has decided to reform its coastal management system and take the approach of integrated management. In 1994, the Chinese government worked together with the UNDP to build an Integrated Coastal Management demonstration zone in Xiamen. Taking advantage of this move, Xiamen reformed its coastal management system according to its own conditions.

In 1995, Xiamen founded the Marine Management Coordination Group led by the Senior Deputy Mayer as a starting point for switching from an industry-based management to an integrated one. In 1996, the Coordination Group set up a committee of marine experts to provide science-based counseling in the field of marine development, utilization, management and protection. On this basis, the Xiamen Marine Management Office was officially established in February 1997 as a standing agency in charge of integrated coastal management under the municipal government. The office holds a monthly meeting to coordinate coastal management departments on major issues that no one department would like to or could handle individually. It has also clarified the responsibilities and duties of each department and assumed responsibility for areas where no one department is in charge. Now, there is no "fight over the sea" between departments in Xiamen with its basic framework of integrated management and coordination well established and overall management in good shape. This has made Xiamen a pioneer among coastal cities in China.

※ Launching a series of conservation and restoration programs

It is fair to say that after years of exploration, Xiamen's ICM system has developed significantly. Xiamen leads the nation in improving marine ecology and rehabilitation, among which sea-dredging and causeway reconstruction are two notable moves. In 1950s, unable to build cross sea bridges due to technical limits, Xiamen built seven causeways, for example, Gaoji Causeway, Maluan Causeway, and Yundang Causeway, in order to meet the needs of local people and economic development. The construction of the causeways did promote economic growth, but harmed the marine ecosystem. Take Gaoji Causeway as an example. Because the eastern and western sea areas were separated artificially, there was no longer a free exchange and circulation of seawater, and the garbage and silt flowing into the sea accumulated, causing the inner bay seawater to become muddy and smelly. This also weakened Xiamen's hydrodynamics, decreasing tidal volumes, and at the same time, increasing sediment. Not only the shipping business and coastal tourism were badly hit, and the development of coastal tourism, but Xiamen's marine ecology was directly threatened.

Xiamen launched treatment of the western sea areas in 2002 and the eastern sea areas in 2006. The closure of many aquaculture farms takes the clean water back. In September 2008, the third phase of dredging in the sea area was launched and took about 10 years to complete. The municipal government spent tens of billions of yuan and then balanced the cost from resultant economic activities. In order to protect the marine ecosystem, a large-scale dredging project was initiated, which was the first of its scale in the country, and also the largest marine ecological restoration project in Xiamen's history. The opening and renovation of the Gaoji Causeway and Jixing Causeway and the west sea dredging project have raised the net water volume transported from the East Sea to the West Sea to 71 million cubic meters. The hydrodynamic conditions of the east and west seas have been greatly enhanced, thus contributing to the conservation of marine resources at the Xiamen Port and the overall improvement of its marine environment. In 2006 and 2009, Xiamen won awards for "Outstanding Achievement in Integrated Management of Coastal Areas in East Asia" and for "Outstanding Achievement in Sustainable Development in the East Asian Coastal Zone" respectively from the Regional Organization of PEMSEA (Partnerships in Environmental Management for the Seas of East Asia). In February 2013, Xiamen was listed by the State Oceanic Administration of China as one of the first group of demonstration areas for its achievement in developing a marine "ecological civilization". The appearance of the Chinese White Dolphin in the sea around Xiamen is the best example.

The Chinese White Dolphin and Lancelet (Amphioxus) are two rare species closely related to the image of the city of Xiamen. In 2011, the Management Office of the Chinese White Dolphin and Lancelet Conservation Area was officially established. Almost at the same time in July, 2011, the first phase of the Chinese White Dolphin Rescue and Breeding Base Program and the Chinese White Dolphin Museum on Xiamen Huoshaoyu Island were put into operation after two years of preparation and construction. The base will be responsible for carrying out work in areas such as rescue and relief, domestication, breeding and acoustic research related to dolphins, especially the Chinese White Dolphin. Thanks to the protection and management efforts in recent years, the number of Chinese White Dolphins, the focus of protection in the conservation area, has stabilized. The probability of seeing dolphins in large groups has increased. In 2011, groups of Chinese White Dolphins (10 to 20 each time) were seen swimming on three occasions in the western waters. In 2013, pictures of a large group of dolphins jumping in the sea of Xiamen were captured. According to the staff of the Xiamen Marine Rare Species Conservation Area, this was the first time in more than 20 years that more than 30 Chinese White Dolphins had been observed together in Xiamen's waters. Moreover, we find that the Chinese White Dolphin has appeared more frequently in recent years in the Xiamen sea areas. Considering the degradation of biological resources sea, we have carried out the artificial breeding and release of marine organisms for many years and spent millions of RMB in releasing fish fry of both commercial and rare species. After several years of artificial breeding and release, the biological resources in Xiamen's waters have increased significantly.

It is worth noting that dolphins only live in healthy sea waters, so the number of Chinese White Dolphins can be an index to determine the ecological health of a certain sea area. Once the ecological health of the sea is restored, we can expect the number of dolphins to increase. The spotting of a group of 30 dolphins swimming together is a sign of improved ecological conditions of Xiamen's sea area.

Xiamen has restored its marine ecosystem step by step over the past several decades. It has carried out multiple large-scale projects, such as the renovation of sea gulfs and opening of Causeways, sea dredging, and restoration of beach and mangrove forests, and the hard work has paid off. The booming tourism in Xiamen is a case in point.

Speaking of coastal tourism, people often focus on three "S"s : sea, sunshine and sand beach. Covering a sea area of 390 square kilometers, Xiamen has a unique, winding coastal beach line that spreads for 226 kilometers. However, it is rarely known that the unforgettable high quality beach that has attracted so many tourists only covers 15% of the total area, and is the most valuable resource along the coastline.

Beaches have always been considered an important feature of a coastal city, their quality affecting the city's image and development of tourism. As early as 2007, the positioning of Xiamen as both a coastal tourist site and a commercial port have led to a series of protection and restoration projects that have been customized to different types of beaches. Xiamen Kuanyinshan Beach became the first one along the eastern coastline to be renovated. The 1.5-kilometer beach, spreading from Xiangshan to Changweijiao underwent cleaning, sand pavement and extension of drainage pipes, which have helped transform it into a fancy area, where the city's most important cultural and sports activities are held. Many tourists come here to enjoy its beautiful "bathing beach" and "vocational village". Based on the success at Kuanyinshan, the Xiamen Conference and Exhibition Center and its surrounding area that covers 300,000 square meters was also incorporated into the overall renovation plan in 2012. But this time, apart from renovations, new designs were also added such as a Beach-Plant Area and a culture theme square.

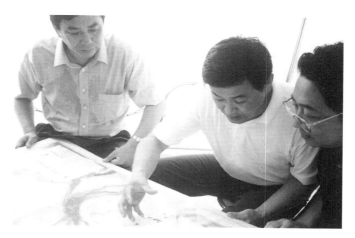

1/2. The opening ceremony of cleaning siltation in waters around the Xiamen Bridge on August 30, 2002

It is true that restoring a beach requires more than simply recovering its old shape, but also perfecting the ocean dynamics to ensure that the sand will not simply erode back into the ocean. At the same time, visual features and functions should be enhanced. The 2013 renovation plan of the Tianquan Bay Beach, tailored to the needs of the Huandao Road, is a good example. In this area, an artificial pebble beach was introduced to replace the previous fragile and chaotic beach with yellow sand after a year's renovation work. Neat, beautiful, and refreshing, the pebble beach has become an iconic site along the Huandao Road. By now, the area of artificial beach has reached 1 million square meters. Past experience demonstrates that renovation of a coastline not only improves scenery, but also sets up a solid foundation for an emerging ocean industry characterized by coastal tourism. Xiamen's shining example has become a model for multiple cities in China and has been introduced to coastal cities in Hebei, Guangdong and Hainan provinces. It has provided valuable experience and technical support for renovation and sustainable development of beaches.

▲ The swimming race to cross the Xiajin Strait, held in Yefengzhai, Xiamen, every July, attracts many swimmers from both sides of the Strait. (Photo / Ouyang Shushun)

※ Managing Maluan Bay in accordance with the principle of integrated management of land and marine, river and sea

The ocean needs protection more than pure management! Through a series of protection programs, such as ecological restoration as well as stock breeding and releasing, the sea around Xiamen has become not only beautiful, but also lively. As a famous tourist city on the coast of the Southeast China, Xiamen lives on the sea, flourishes by the sea, and thrives together with the sea.

The sea has nurtured the people of Xiamen for thousands of years. In modern times, with the acceleration of social and economic development, the Xiamen sea begins to show signs of "exhaustion", such as increased sedimentation, reduced tidal volume and coastal damage, which seriously restricts the development of Xiamen from an island city into a coastal city. We have implemented coastal governance, improved the quality of waters and coastal ecology, expanded city space, promoted economic development, and advanced the quality of development. Since the 1990s, Xiamen has made remarkable achievements in its coastal management. For example, through the integrated management of Xinglin Bay, we built the Yuanboyuan Expo on the water, achieving flood control, drainage, and environmental protection and creating ecological harmony.

Now, the integrated management of Maluan Bay, Xiamen's "last bay", has proceeded like a raging fire. In recent years, by integrating land and marine administration, we have explored diverse approaches to reduce the flow of pollutants from land to the sea. Based on the experienced of Xinglin Bay and Wuyuan Bay, Maluan Bay has made new breakthroughs in terms of the level and scope of regulation. Taking the approach of integrated management of "both the high mountains and the sea", we no longer focus only on the bay, but also over 10 square kilometers of river basin leading into Maluan Bay.

In the case of Maluan Bay, the success of coordinating land and marine administration hinges on the fact that all the rivers flowing into Maluan Bay are within the territory of Xiamen, and no other administrative region is involved. That's why we are able to explore the integrated management approach at Maluan Bay and it is a key project underway.

Caught between Tianzhu Mountain and Caijianwei Mountain in the northwest of Xiamen, and connecting Haicang and Jimei districts, Maluan Bay is considered an important pass-way between Xiamen and Zhangzhou, and an axis of the Golden Bay. The overall city planning has designated Maluan Baym not only a secondary city center of Xiamen for its special geographical location but also a strategic step in creating an "alliance of cities".

Before the 1960s, endowed with exuberant mangroves, clean water, white egrets and sailing boasts, Maluan Bay was known as one of the most beautiful bays in Xiamen. However, its landscape changed dramatically after the completion of the causeway. The Maluan Causeway used to be the "guardian spirit" for Xiamen's salt industry, agriculture and aquaculture, but despite the important role it played in the development of these industries, the causeway isolated the bay from the outer sea. Without any water exchange, it basically became a pond of stagnant water. With the passage of time and economic development, the negative impact of the causeway became more pronounced. The results were decreased tidal volume in the inner bay, sedimentation and ecological deterioration. Opening and renovating the causeway became a musty.

After 10 years' preparation, the Maluan Bay Causeway Opening Project finally began in 2005s to promote water circulation, increase tidal volume, and improve water quality.

In fact, the idea of managing Maluan Bay was first brought up in as early as the 1990s. Since 2003, related hydrodynamic studies with physical and numerical models have been launched.

The bay area, sluice setting, coastline layout, water exchange and water level of the landscape have been identified and relevant reviews have been conducted. The Maluan Bay project was the most studied one by expert panels who had set numerical and physical models for several rounds. But for a variety of reasons, the whole project lasted for more than 10 years and large-scale construction did not begin until last year.

It is worth noting that the integrated management and development of Maluan Bay is a system-wide project. Revitalizing the Maluan Bay ecosystem is not only the first step, but also the premise and the foundation of the whole project. The Maluan Bay Causeway Opening Project also involves dredging, coastline construction, wetland and river system governance, and ecological restoration of the waters and the surrounding area. We also need to improve the Maluan Bay hydrodynamic conditions and its surrounding environment, increase the tidal volume of Xiamen's western sea areas, manage aquaculture ponds and low-lying lands in the upstream of the planned waters and to govern the river system. Only by doing so can we realize integrated management and the complete restoration of the Maluan Bay environment.

At present, cutter suction dredgers are in place, carrying out dredging in a tight and orderly manner. The work proceeds continuously at a large scale. By the middle of December 2016, 500 million yuan had been expended on the project. Overall silting and site rehabilitation will be completed by the end of 2017. At the same time, we have conducted research and constructed mathematical models related to water gauge gate's dispatch and power flow of the Maluan Bay landscape, laying the foundation for coastline construction. We plan to solicit proposal for coastline construction, complete preparations for the Maluan Bay Ecological Wetland Renovation, and gradually start construction according to the progress of ecological restoration and the new town development in 2017, adding new beauty to the Maluan Bay ecological restoration.

A year ago, there were nothing but shoals, but now the Maluan Bay has become a spotlight, attracting people's attention. Throughout this area, the horn of development has been blown—projects for improving local livelihoods have been completed and ecological restoration is

1. Before rehabilitation, the route through Makui Bay was crowded with fish rafts, cages and sea bats farming.
2. A panoramic modeling of Makui Bay
3. The original view of waters in Maqu Bay

fully under way. The new town at the Maluan Bay is changing dramatically. At present, as a pilot program, phase one of Maluan Bay Ecological Restoration Project has already started. It mainly focuses on water dredging and ecological rehabilitation, with an aim to rehabilitates an area of 17 square kilometers area, expand coastline by 14 kilometers, and increase tidal volume by 30,000,000 cubic meters. The sea area after dredging will be expanded to 8 square kilometers.

According to our plan, the total area after renovation will be about 37 square kilometers, with a planned population of around 400,000. With Xinyang Bridge at its eastern end, Fulian Road at the western end, Wengjiao Road at the Southern end and Haixiang Avenue at the northern end, Maluan Bay will serve as a secondary center of Xiamen, as well as an important gateway for city integration (between Xiamen and Zhangzhou).

However, from the perspective of integrated coastal management, the key in the management of the bay area is to control the water, in other words, maintain water quality. At present, we plans separate fresh water at the inner bay from salt water at the outer bay. Six square kilometers of the outer bay exchanges water with the western sea and one square kilometer of the inner bay exchanges water with estuaries and is supplemented by recycled water plants. But during the project, through repeated studies, we found that the inner bay in the dry season mainly relies on the supplemental water from recycled water plants, but in the short term, the water plants only release 25 thousand tons of water per day, despite the projected amount of 137 thousand tons per day in the long term. In other words, water exchange has been insufficient in the recent low water period.

Evidence shows that the water quality of the inner bay is likely to deteriorate, which in turn will harm the overall image of the new town. Therefore, improving the water quality of the bay area is a goal of the next stage of our work. Currently, we are conducting a systematic research on the water quality at both the inner and outer bay to study the potential risks and precautionary measures for Maluan Bay in case of the dry season, insufficient water supply from recycled water plants, and an incomplete prevention of siltation during different construction stages of the new town.

The sea needs management but also protection. To sustain its blue color, Xiamen has stepped up its efforts in restoring marine ecology in a multifaceted and multi-layered way. Over the past two decades, Xiamen's ICM has gradually shifted from managing space utilization to ecological protection and resource conservation. It has also become the leader in upgrading the marine economy from relying on primary industry to secondary industry and further to tertiary industry. The overall system for utilizing sea resources has been optimized and upgraded, bringing benefits to the marine environment. We are all delighted to see more dolphins choose to live here.

The Success of Yundang Lake Case Lies in Integrated Management

Zhang Bin

Southern Chinese often describe the northerners as "landlubbers". Zhang Bin, whose ancestral home was in the north but birth place was Xiamen, learned to swim skillfully in the sea of Xiamen. "During the 'Cultural Revolution', I lived at the Port of Xiamen for ten years, and spent much time in the sea. " In his opinion, there must be something wrong if any Xiamen local shows no affection for the sea.

But his attachment to the sea is far beyond that.

In 1993, when the Xiamen government started building Egret Islet on Yundang Lake, Zhang Bin was transferred from a public transportation institution to Egret Islet Construction and Development Company, thus forming his ties with the construction and management of this lake. Later, the integration of the company and the Yundang Lake Management Office enabled Zhang to be more directly and deeply involved in the flood prevention, sewage treatment and greenery maintenance for Yundang Lake. In 2007, he was transferred to the Yuanboyuan Expo to take charge of its construction and management. Nevertheless, Zhang was still closely connected with Yundang Lake in his practical work since the Garden Expo also belongs to the municipal garden system. In 2011, Zhang went back to Yundang Lake Management Center to preside over the 5th integrated management of this lake, the strongest effort ever. In 2015, he was transferred from the center to Xiamen Environment and Sanitation Management Service. "It is fair to say I have participated in, directly or indirectly, the Yundang Lake management from 1993 to 2015. " told Zhang Bin. All Xiamen residents are fond of this lake. As a worker who has participated in the lake clean-up, he has devoted endless efforts. That's why he would patiently explain to those who raised questions about the lake management and immediately stop uncivilized behaviors that occurred near the lake.

"Over these years, I have heard people complain that the lake still gives out odor and requires dredging from now and then despite the large amount of money spent on its management. I just want to tell them: the lake itself has nothing to do with the smell; as long as the lake exists, our dredging work will never stop! If the basin is left unregulated and population grows uncontrolled, the lake management has to take place once every few years! The Egret Islet Park is more than a park; it is an important facility for flood prevention for one third of Xiamen's territory, and is as important to the city's environment as kidney is to human body. To protect Yundang Lake—our city's new living room, Egret Islet, and Xiamen Island's environment, we must protect the lake as if we protect our own eyes. "

※ Lake's destiny determined by geographical factors

Speaking of Yundang Lake, one should start with its history.

Yundang Lake was once a harbor. More than a dozen years ago, the Yundang Lake Management Office wrote a book *From Yundang Harbor to Yundang Lake*, which depicts the lake's transformation. While editing this book, we looked into numerous documents, consulted historian Mr. Hong Buren among other pundits and interviewed old fishermen living many years at Yundang Harbor. The oldest existing map of Yundang Lake shows the lake occupied over 16 square kilometers in the early Qing Dynasty but only around 10 square kilometers when the Opium War broke out in 1840. The original boundary of the lake stretched to Jiangtou Park, a former large port, from where in ancient times large ships could directly sail to the sea.

What was behind the shrinking? Because Yundang Lake was a lagoon, the rain that fell on

Introduction to the speaker

Zhang Bin, deputy director of Xiamen Environment and Sanitation Department, once served as the director-general of Yundang Lake Management Center and CPC branch secretary. He has been engaged in comprehensive rehabilitation of Yundang Lake since 1993, and has been in charge of the fifth phase of comprehensive renovation of Yundang Lake.

▼ Map of Xiamen City in the 19th Year of Daoguang, Qing Dynasty(1839). (The map was repainted by Zeng Huanhui in 1988)

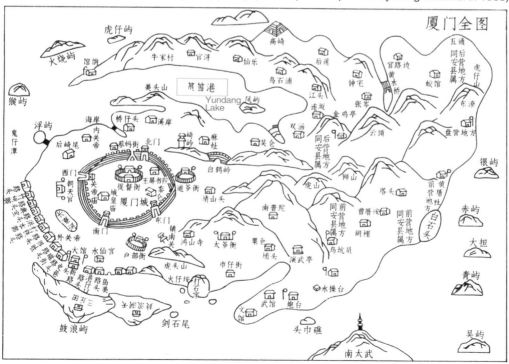

one third of the catchment area on Xiamen Island flew into the sea through this lake. When it rained, the rainwater would carry the sand on the mountains and basin into the lake. In history, plenty of sand-trapping mangrove trees were planted along the seashore of Yundang Harbor, which meant the intertidal zone could grow instead of being washed away. In the case of a seaside lagoon, the beach would extend into the sea; in the case of an inland lake, the intertidal zone would stretch to the center of the lake, thus reducing the water body. Yundang Lake was the latter case.

Turned from an earthquake fault zone, Yundang Lake was a harbor that resembled a deep pot, whose bottom was much lower than the estuary. For a long time, the muddy water down from the mountains filled the lake. Sand and soil were also carried to the lake with the rising tides from the western part of the sea. When the tides rose, the seawater stayed in the lake for a while, and when the tides fell, the sand deposited instead of returning to the sea. Therefore, Yundang Lake was described as a pot, whose bottom would trap anything that came inside. As time went on, silt was slowly sedimenting in Yundang Lake.

Some citizens are often confused why Yundang Lake is always being dredged. We tell them that dredging is the only solution. In particular, at the early stage of urban development, mountains were dug into for soil, which brought large amount of sand and mud into the lake. Every 5 or 6 years, we need to clear over 1 million cubic meters of mud. Therefore, the entrances of floodways must be dredged every year and large-scale dredging be done once every few years. Otherwise, the storage capacity would be fall and the mud would stink and affect water quality.

Yundang Lake does not produce soil itself, but the surrounding basins keeps carrying soil into the lake. Due to the lake's poor ability to exchange water with the sea, more sand would come in and flow out, which means the lake would become smaller. Dredging and other human interventions might slow its shrinkage and prolong its life, while man-caused disruptions, such as sand and sewage from excessive economic activities and by residents on its shore, would shrink the lake until it disappears. I always believe that anything has got its own life cycle. Lakes are no exception. Some lakes are doomed to disappear. Any attempt to save them at best help to delay that inevitability—the natural order of things.

But none of us would want this to happen to Yundang Lake.

In the early days, Xiamen, Yundang Harbor in particular, was thinly populated. The place across the Xiamen Railway Station was once called Wucang. Over 300 years ago, a Wu family started to fill the intertidal zone with soil in order to plant crops; this was among the events that marked the beginning of Yundang Harbor development.

▼ Before 1970s, the western part of Xiamen Island, Yundang Port and the Xiamen–Gubei Strait were inter-connected geographically. (Photo provided by Bai Hua)

◄ Weeds in Yundang Lake before comprehensive rehabilitation

◄ Garbage dumps in Yundang Lake before comprehensive rehabilitation

In fact, the official history of Xiamen began when Yundang Lake was bustling with human activities. The year when Zheng Chenggong led his army to Xiamen witnessed the largest migration of the early days to the city, which largely impacted Yundang Harbor. This was not accidental: the geographic of Xiamen Island meant that its development was bound to play out around Yundang Harbor. When productivity was low, people preferred livable places on the southwest of Xiamen Island to avoid natural disasters including typhoon, and areas surrounding Yundang Harbor became the first choice. For example, "Chen and Xue", two richest families in the early history of Xiamen, chose to settle near the harbor. The development of the eastern part of the island only began recently.

An increased population in the surrounding areas posed a threat to Yundang Harbor far earlier

◢ The revetment of Yundang Lake after rehabilitation in 1988

◢ The first large-scale effort to clean up siltation, which was used for land reclamation, the later Egrets Park and Egrets Island, in 1988

than we thought.

In 1937, a number of "national capitalists" in Xiamen were planning to build a fishery at the harbor to for high-end fishes. They invited the then principal of Jimei Fisheries School to conduct research, which concluded the water of the harbor was already heavily polluted. What were the sources of pollution? First, dozens of nearby creeks carried lots of garbage to the harbor on rainy days. Second, most excrement in the city was dumped in the harbor, forming a big cesspit on the West Hexiang Road nearby; this situation remained unchanged until the reform and opening up. Very few Xiamen households were equipped with toilets, so they mainly used buckets when practicing defecation. Every early morning, sanitation workers, pushing their flatbed tricycles, collected human feces from door to door and dumped them in the cesspit. Afterwards, farmers from Longhai got the excrement and used it as fertilizer back home. The repeated transportation of such waste hugely damaged the water quality of Yundang Harbor.

But why did the pollution seem less severe at that time? First, despite the severe pollution, the city's population was still small, only dozens of thousands before the foundation of the People's Republic of China. Second, the harbor could exchange water fairly well with the sea, thus dispelling much of the odor. Third, the sources of the pollution mainly came from daily life, as there were barely no industries developed in Xiamen at that time. Despite all these, the lake was already suffering from severe pollution.

With urban development and population growth, after the foundation of the PRC, Xiamen's industries played out along Xiahe Road. Almost all factories were situated near the road, and

most citizens lived there as well. Today, the names of many roads in Xiamen bear the character "Xi" (which literally means creek in Chinese, since those roads used to be creeks of various sizes, with some directly linked to Yundang Harbor. Though not a big flow on sunny days, when it rained, those creeks carried a large amount of water flowing into the harbor. Larger population and more factories meant more sewage and other waste into the creeks, and then to the harbor, thus damaging the water quality in the lake.

The issue of Yundang Lake pollution culminated in the 1970s.

At a time of full combat readiness, Xiamen, the frontline, mainly relied on other places for food. Having the coastal wall as its only transportation channel, Xiamen would face food shortage once the wall were blown up. The "Revolution Committee" demanded that Xiamen be self-sufficient in crops within 3 years.

▼ Yundang Lake is known as the "living room" of Xiamen. The purpose of the comprehensive improvement project of Yundang Lake is to further improve water quality and enhance landscape effects of the surrounding region. (Photo / Wang Huoyan)

How could Xiamen, a coastal city with little arable land, solve this problem? Some suggested enclosing Yundang Harbor and reclaiming land from the sea.

But the suggested solution backfired.

The land reclamation plan was to build a causeway that stretched from the current North Siming Road to Dongdu Port, with a small opening left for drainage and flood prevention. Before the plan was carried out, the lake, despite the inflow of much polluted water, was still able to exchange its water with the sea, so the water was still flowing. After the land reclamation, only dirty water flew into the lake, severely damaging the water quality. While Xiamen was pursuing rapid industrial development, a number of factories sprung up, such as printing and dyeing factories, leather factories, chemical factories, and sign-making factories, as well as Zhongshan Hospital, discharging huge volumes of wastewater into the lake. After the land reclamation in 1970, the output of fish and prawns plummeted the following year—the most obvious sign of the worsening water quality. Had not it been for the project, the lake could have provided several hundreds of tons of sea food every year for local residents.

What followed the output drop was a stinky lake and a deteriorating eco-environment. How smelly could the lake be? There was a vivid description: all fishes and prawns in the lake died out except one fish—the most pollution-enduring climbing perches; all plants in the lake died out except the most pollution-enduring algae; the bank was awash with rats, cockroaches and flies. The entire lake turned into a septic tank. Vehicles passing the road on the West Dyke all had their windows closed. I used to work in the public transportation institution near Yundang Lake. When the colleagues finished work and walked along Xiahe Road near the current Ophthalmology Hospital in the evening, nobody dared to talk, as they might accidentally take in about ten mosquitoes once they opened their mouth.

※ Five phases of integrated management

The situation lasted from the early 1970s to the end of 1980s. During that period, Xiamen Daily once featured a cartoon called "Watching TV under a Mosquito Net", which vividly depicted a family watching TV programs under a mosquito net, surrounded by a dozen burning mosquito-repellent incenses. This cartoon caused great response among the public.

Meanwhile, both the municipal CPC committee and government attached great importance to the Yundang Harbor issue. When I visited the former mayor Zou Erjun around 2000, he told me, "When I was the mayor, the annual disposable fund for the whole city was mere 8 million yuan, covering salaries for doctors, teachers and civil servants as well as the expenditure on public facilities." His remarks showed how determined the local authorities were to harness the pollution at Yundang Harbor, despite such a limited budget. Since the latter half of the 1970s, the municipal government had been studying how to control the lake pollution. They once applied to build a sewage treatment plant, and held seminars where experts were invited to explore specific measures.

In the early 1980s, the government had already taken several measures to curb pollution, such as suspending operation of heavily-polluting companies, a rather major move. The government subsidized those closed firms despite the financial difficulties, reflecting the leaders' boldness.

One year after the clean-up, these efforts took effect, with almost 70 percent of the lake pollution sources stopped. But more should be done. Although less sewage and sand was brought into the lake, the waste water already at the bottom of the lake was still fermenting, giving out a foul smell; the lake was stagnant due to insufficient water exchange. The odor was so bad that clean-up was a must.

At that time, Yundang Lake was described as a trauma of Xiamen.

The massive integrated management of Yundang Lake started 1988. On April 12, 1988, a significant review, "Do Not Meet the Public Until You Clean Up Yundang Lake", was published on Xiamen Daily, reflecting the public's expectation and the government's resolution. Articles of this kind were rarely seen on CPC newspapers. In September of the same year, the Standing Committee of the People's Congress in Xiamen approved a proposal for speeding up the integrated management of Yundang Lake, and the municipal government conference passed a decision that a hefty fund of RMB 20 million would be allocated each year for lake clean-up in the following 3 years. In 1989, the first phase of integrated management officially kicked off.

Between 1989 and 1992, a guideline was set for the first round of integrated management—intercepting and disposing of pollution sources, getting the water running, dredging the silt and building the bank, and improving the environment.

The first step was intercepting and disposing of pollution sources, which consisted of two parts—intercepting the sewage and disposing of it in sewage treatment plants. To this end, the city built a sewage treatment plant with a daily secondary-treatment capacity of 37,000

tons (45,000 tons in the actual operation capacity) and a daily primary-treatment capacity of 134,000 tons, together with a network of sewage interception pipelines that stretched 24km.

The second step was getting the water running, or diverting the seawater to the lake. Even though the pollution sources were intercepted, the lake was still stinking if the water was stagnant. As the old saying goes, "flowing water stays fresh". We built a stretch of diversion dyke at the entrance of West Dyke; taking advantage of the gap between the ebbs and flows of the sea, we drew the seawater to the lake when the tide rose. The natural flow of the seawater alone could not lead to the water exchange in the lake. With a small water inlet, the floodgate could be closed when the water level in the lake rose to a certain level, since the water level of the open sea was higher than the shoreline of the lake and uninterrupted inflow of water would drown South Hubin Road.

1 | 2

1. The construction team was busy installing a pumping station. After installation and commissioning, the sewage around Songbai Lake could be intercepted. (photo / Liu Donghua)
2. The waters of Crescent River in Jiangtou Park flow into the downstream of Songbai Lake and exchange waters with each other. The water in the Songbai Lake will flow into the Wuhu Lake in order to achieve water exchange. (photo / Liu Donghua)

Yundang Lake was misunderstood by many as a mere park. They only cared about the scenery and smell of the lake, but knew little about its importance to our city as a flood-prevention and drainage facility, which, in fact, was its primary function. Over the past 30 years, the stability of the lake guaranteed the security of the central Xiamen, a result of the efforts made by generations of cadres and employees at Yundang Lake Management (Office) Center.

The diversion dyke, aimed to get the lake water running, successfully helped draw the seawater into the lake, and then to the main canal on the east part of the lake, thus making the lake water circulate.

The third step was dredging the silt and building the bank. The entire lake was suffering from a heavy layer of sand, making dredging a necessary move. Where should the dredged sand go? It was used to build Egret Islet. At this phase, a total of 3.2 million cubic meters of silt was dredged, adding around 700,000 square meters to the water surface, with a 10-kilometer-long revetment and an around 700,000-square-meter bank taking shape.

The fourth step was improving the environment. But little was done in this regard due to limited government funding, leaving behind only a patchwork of greenery dotting the shoreline.

It is no exaggeration to say that the first phase of management immediately and fundamentally changed the lake, especially its water quality. The water became less smelly and much more clear, even allowing dragon boat and motorboat competitions to be held in the lake.

There was a funny story related to the motorboat competition at Yundang Lake: in 1995, when the Formula 1 Powerboat World Championship was to be held in China, General Secretary of the International Motor Sports Association (FIM) came to China to choose venues for holding this event. Xiamen was not among the 5 cities suggested by the General Administration of Sport of China and the Federation of Automobile and Motorcycle Sports of People's Republic of China. Coincidently, FIM General Secretary arrived in Xiamen after his international flight. During his city tour in the company of officials of General Administration of Sport, when roaming Egret Islet, he chose Yundang Lake as the competition site at his first sight of it. Those Chinese officials told him the other 5 cities were better, but the general secretary could not believe that there should be a better place in China.

At that time, Yundang Lake, with fewer skyscrapers, a clear skyline and a broad horizon, was more beautiful than in these days. According to F1's requirements, the competition should be held in the downtown area to maximize its publicity, but it was difficult to find a sea area in the middle of the city throughout the world. Knowing the competition tracks on Yundang Lake were not up to code in length, the General Secretary decided to change the rules. The water environment of the lake was so impressive that a world competition would rather change its rules than change the site, which showed how successful the first phase of integrated management was.

However, after the completion of the first-phase management, new problems popped up right away due to rapid population growth. In 1989, the development of Xiamen Special

Economic Zone took off. The original sewage treatment amount of 20,000 to 30,000 tons per day in Yundang Lake used to be sufficient, but a surge in the urban population increased the sewage to over a hundred thousand tons, and even more, forcing the urban sewage disposal ability to upgrade. On the other side, considering the city's urbanization degree, the lake's flood-prevention ability was far from enough to ensure urban safety. Flood control had been relying on natural floodgates in this city. Drainage would weaken during neap tide, and the water could hardly be drained out during high tide, both making it difficult to clear room for storage capacity. This situation could affect the safety of the area neighboring the lake and the 37-square-kilometer urban area in the basin. A growing number of houses built around the lake, made flood prevention even more urgent.

As a result, the second phase of integrated management started in 1993.

The primary step of the second phase was to deal with the ever-increasing daily sewage, which skyrocketed from 120,000 cubic meters per day in 1989 to 200,000 cubic meters per day in 1993. About a half of the effluent could be intercepted and treated, but the remaining half would flow directly into the lake, forming a new pollution source. To overcome this, the second sewage treatment plant in Xiamen was built with a daily primary-treatment capacity of 100,000 cubic meters (150,000 cubic meters in the actual operation capacity), and a daily deep-sea drainage capacity of 250,000 cubic meters.

The second step was to prevent and control flood. We built a pumping station with a drainage capacity of 40 cubic meters per second, lowering the frequency of the city's possible flood disasters to "once in five decades" from "once in a decade". When we could not take advantage of the lower water level of the open sea to drain water, we could actively pump water from the lake to prevent flood.

The third step was to increase the revetment. Part of the revetment was roughly built at the first phase, so we built a 14 km revetment and 27 km lake-surrounding green waterfront at the second phase.

The fourth step was to expand the green area. At the second phase, we set up a green area of 460,000 square meters and built 16 small squares, with more leisure functions.

The fifth step was to complete the project of diverting water from east to west. Upstream of Yundang Lake sat Songbai Lake and Tiandi Lake, which were stinky since there was no fresh water flowing in and the water was stagnant when it was not raining. On rainy days, the effluent in those two lakes flew to Yundang Lake. The water overflowing from the two lakes would drift to Yundang Lake, so they formed a pollution source to Yundang Lake. In this regard, we built a sewage lift and pump station to pump the clean seawater into Tiandi Lake and let it flow downstream, increasing the water flow in Tiandi Lake and Songbai Lake. With dozens of thousands of cubic meters of seawater being pumped every day, the water flow, though still insufficient, was better than nothing.

In addition, we added another trunk sewer on the south bank since the first sewer could no longer meet our needs. We built a 14m-high Egret Goddess statue, which was a landmark of Xiamen, a large music fountain that was Xiamen's first, and a 45m-high stature of "an egret spreading its wings" to show the title of "National Model City" granted to Xiamen. For the first time, we started our project for creating night scenery in the lake area.

When the second phase of integrated management was done, the lake environment was hugely

▼ In November 1997, the International Motorboat Alliance selected Yundang Lake as the venue of Gaoli F1 Motorboat Championship and achieved a complete success.

improved. Both the cultural environment and the natural environment, especially the green area, were better compared to the first phase. The flood prevention capacity was also enhanced.

But some problems were left unsolved after two phases of management—dozens of thousands of tons of sewage in Songbai Lake and Tiandi Lake could not be intercepted. On top of that, since the urban development, a host of creeks were turned into floodways, some of which had been intercepted and thus deposited much sand and silt. The diversion dyke of Yundang Lake, made of soil, was heavily damaged by the scour of the seawater over the last ten-odd years. Once the dyke broke down, the water circulation project would fail; so we must consolidate the diversion dyke.

Therefore, we started the third phase of integrated management in 2005. First, we consolidated the diversion dyke, which was made of silt in 1988 and now broken after absorbing and draining the tidewater for about two decades. In flood seasons, the dyke became very precarious, which could trigger flood and threaten the normal drainage and absorption of the lake. So, we transformed the around-2km-long internal and external diversion dyke into solid, endurable concrete stone posts.

The second step was to dredge the lake. We dredged over 1.3 million cubic meters of silt. We started in 1989 and performed small dredging tasks every year. In 2005, we undertook the largest dredging project. We used a cutter suction dredge to suck up the soil from the bottom of the lake, and transported the muddy water through a 14km-long pipeline to an abandoned shrimp pond in Hecuo, where we deposited the soil and drained the clean seawater into the sea. The workload was so heavy that once our work was done we scrapped the pipeline, which suffered serious wear and tear due to fast sand flow. During the whole process, we stayed on high alert and assigned staff on patrol for months, worrying that the pipeline, once broken, would hurt passers-by with the spurted sand.

The third step was to explore ecological restorations. To explore relevant technologies and measures urgently needed by the clean-up of Yundang Lake, we conducted studies of integrated management of salt water lagoon in the sea front and algal bloom in the lagoon, smooth operation using patented technology, and lake management using advanced plants such as mangrove trees and alkaline-tolerant plants. Those studies provided important scientific foundation for dredging and intercepting pollution sources, smooth operation of the water body and a new phase of overall planning for Yundang Lake. Among them, the study of "Smooth Operation of the Water Body" won the second prize of the Science and Technology Award in Fujian Province.

After three phases of integrated management, the main problems of the lake had been alleviated. To consolidate achievements and better the lake environment, we diverted our attention to basin management. That was how the fourth phase of management unfolded.

The major tasks at the fourth phase included: first, building a sewage lift and pump station with a daily lifting capacity of 50,000 tons at Jiangtou Park to pump the effluent into the sewage pump station on Lvling Road, as a way to lower the pressure on the trunk sewers surrounding Yundang Lake; second, intercepting sewage from 6 outfalls near Tiandi Lake, which would reduce the sewage flowing into Yundang Lake by nearly 10,000 tons a day; third, blocking the release of effluent from floodways near Yundang Lake, which would decrease nearly 12,500 tons in the daily amount of sewage flowing into Yundang Lake; fourth, deodorizing stinky floodways.

After the fourth phase, the water quality of the lake was much improved, with the density of mineral nitrogen down from 2.5mg/L to 1.15mg/L and the density of inorganic phosphorus down from 0.15mg/L to 0.06mg/L.

Those four phases of integrated management of Yundang Lake achieved remarkable results. With the social and economic development, however, new problems kept emerging: due to a higher floor area ratio of the city and a higher population, the amount of sewage produced in the basin exceeded the planned amount, so the trunk sewers, failing to meet the needs, oozed some effluent into the lake; the sewage from some floodways near the upstream Songbai Lake had not been intercepted, directly flowing into Yundang Lake; the drainage system could not thoroughly separate sewage and rainwater, making some wastewater in the system flow into the lake during rainy seasons; silt and sewage in floodways gave out a strong odor; etc. The fifth phase of integrated management thus got on the agenda.

The fifth phase started in 2010, with the main tasks including:

First, Songbai Lake clean-up. The sewage from all outfalls near this lake should be intercepted and the lake should be completely dredged.

Second, we would built No.1 Binbei Pump Station to pump 100,000 tons of sewage a day to a sewage treatment plant in the east, so as to lower the pressure on Yundang Sewage Treatment Plant. To be more specific, this new pump station was designed to pump over a dozen tons of sewage froms 17th and 18th floodways in the east to the Shizhoutou sewage treatment plant.

Since the sewage sent to the West Dyke Sewage Planthad exceeded its treatment capacity, the intercepted sewage, when being pumped into the pipelines at the plant, could still overflow and drift into Yundang Lake. So we diverted 100,000 tons of sewage to Shizhoutou Plant, reducing the pressure on the trunk sewers flanking Yundang Lake. We added a large trunk sever on the stretch of the south bank near the sewage plant as Yundang Lake was situated on mobile silt, he and the resultant land subsidence might disjoint the original trunk sewers, built according to a low standard, after years of use.

Third, we intercepted sewage flowing from all floodways on the north and south banks and built lift and pump stations at the entrance of the floodways, to prevent sewage from entering the lake at fine weather. We built a pump station at the entrance of each floodway, to stop sewage. We also upgraded the existing pump stations, which ensure no sewage would flow into Yundang Lake on sunny days as long as the sewage disposal plants were not undergoing blackout or repair. We successfully guarded the last pass.

Fourth, we transformed the floodways on the south and north banks of the lake.

Fifth, we dredged the floodways in the basin. Those floodways, which had not been dredged for years, might carry garbage and waste along with the rainwater into Yundang Lake, which would not only pollute the water body, but also harm the lake's capacity to drain flood, posing safety threats.

Sixth, to tackle both symptoms and root causes, we cleaned up and transformed the floodways by separating sewage and rainwater flowing in them and strengthened drainage management. Floodway management was crucial. 20 years ago, due to limited residential area, most of residents would use the balcony as the toilet and kitchen once moving into their new apartment, even if the apartment was newly built. Each balcony was equipped with a rainwater pipeline directly linked to floodways. The result was that the domestic sewage flew into the lake through floodways. In another case, some residential communities did not clean the plugged sewers but connected them to the nearest floodways, so the domestic sewage flew into the lake through floodways. The third case was that some restaurants directly poured sewage and waste into roadside ditches and as cooking fume was not allowed in the air, it was discharged into the floodways, causing pollution inside. The sewage thus entered the lake through the floodways. Also, the sewage incurred by washing septic tanks and cars also drifted into the floodways. The condition of the floodways in the basin directly affected the water body of Yundang Lake, therefore the urgency to clean up.

Seventh, landscaping. The design principle for the greening upgrading was to maximize the building of scenic gardens. We added some paths and boulevards, and transformed the soil. At Yundang Lake, the greening level started low and failed to improve much over the last three decades. The soil was heavily hardened due to salinization, which affected the growth of plants, making the soil transformation necessary.

We chose jasmine, daphne odera, milan and other fragrant species when planting flowers, and this worked well. Meanwhile, we emphasized the aesthetics of plantation. In the past, as long as trees were planted, composition was not a concern. But now, we focused more on the variety of colors. We introduced Tabebuia impetiginosa, cockscomb and wiliwill, among other flower-growing trees, thus adding beauty, color and fragrance to the lake—a dream come true for several generations of locals near Yundang Lake.

We also expanded the lakeside paths and boulevards in the lake area, and installed lampposts for some footpaths in this area. We used permeable bricks for all the paths, creating a nice landscape.

A great number of floodgates built during the integrated management were located on the lawn, which was quite an eyesore. So we went out to hide them. For example, in front of the Egret Islet Hotel, we transformed the floodgate into a southern Fujian style brick wall. An inspection shaft, bulging like a fort, was turned into an ancient style well and around the well we built a square for public recreation. For a stretch of a sewer that had to cross the entrance of the floodway, we built a sightseeing platform when the project was done, perfectly hiding the sewer under the platform.

The effort at this phase significantly improved the landscape of Yundang Lake, which was highly commended by citizens and tourists.

※ Integrated management is an unending process

I believe the integrated management of Yundang Lake is an example of human learning about the ecosystem and respecting nature. This lake has become a demonstration site for Partnerships in Environmental Management for the Seas of East Asia (PEMSEA), and we would share our work experience and learnings from the management projects at the training sessions held twice each year. In my opinion, human being should respect nature instead of damaging it, which is of great importance. Second, if we accidentally do harm to it, we should figure out how to restore it. Third, the restoration should be comprehensive and systematic.

Since the first phase of the integrated management, we have concluded that we should not only intercept sewage, but also dispose of the intercepted sewage; otherwise, our effort would be equal to sewage transportation and secondary pollution once the intercepted sewage goes into the sea—the problem remains unsolved. Once this problem has been addressed, we should enhance the water flow capacity to secure the self-cleaning ability of the water body. And we also need to restore the eco-system. The lake clean-up hinges upon both the water quality and fauna and flora in the water. Following those five phases of work, the eco-system in the lake area has been immensely improved, with several indexes for water quality outperforming the four-level standards for seawater. The lake is now home to over 70 kinds of birds, 43 types of aquatic organisms, 201

◄ Mangroves in Yundang Lake
(photo / Wang Huoyan)

kinds of phytoplankton and over 70 kinds of zooplankton.

In fact, Yundang Lake also affects the city's air quality, ecological environment, and the western sea areas. The water quality of the lake, when it is static, is better than that sea area. The depolluted lake has in turn improved the environment of the sea, at least reducing the pressure on it. And Funding Lake also has many invisible ecological impacts.

The five phases have greatly improved the ecological environment and water quality in the lake, with the clean-up result highly commended by visiting domestic and overseas experts. But there is still room for environment protection and upgrading—more could be done in the basin. How can we properly manage the basin? How should we clean the floodways and make the effects last? How should we completely separate sewage from rainwater and free the floodways from sewage? Those are issues left to be dealt with. With more government funding in the future, I would suggest carrying out primary rainwater treatment.

The eco-environment of Yundang Lake can be further improved, but now it is overcrowded with people. I think we should not treat the lake as a park but as an ecological area, with fewer people, more natural trees, various birds, and more diverse aquatic plants and animals. When the population grows, its ecological functions naturally go down.

We tend to crowd any nice place once we find it. A city should have both lively and quiet places. With too many busy areas, Xiamen should also have some tranquil places for citizens and tourists who prefer quietness. It would be better if we could preserve the ecological functions of the lake and, guided by policies, turn it into a ecological zone for those who love tranquility.

The impressive clean-up results are attributed to the focus and leadership of the municipal government and CPC committee of several sessions, the support of the general public, and the devotion of all the scientists, technicians and workers who participated in the integrated management.

I have left the Yundang Lake Management Center, but I still have suggestions to offer as an old Yundang local: never be content with the status quo. The clean-up has indeed achieved good results, but beware that the lake's ecosystem remains fragile and the responsibility of flood prevention still heavy on our shoulders. We should do more to make the lake into better.

The odor of Yundang Lake has already gone.

But the lessons learned in the hard way should never be forgotten.

Management of Lake Is to Protect the Lung of Xiamen

Lin Xueping

In 1996, after graduating from the Soil Chemistry Department of the Fujian Agriculture and Forestry University, Lin Xueping was sent to the Division of Flower Production of the Xiamen Municipal Government, a neighboring city of her hometown Zhangzhou.

Lin said: "Back then, the Division of Flower Production was very close to Yundang Lake and my dormitory was also nearby. " Since the day when she set foot on Xiamen, her life has always been associated with this lake. Though she does not live around the lake now, her work is still related.

In 2008, the Management Division of Yundang Lake, Management Division of South Lake Park and the Division of Flower Production were merged into the Management Center of Yundang Lake. As Deputy Director of the center, Lin was directly involved in its management. Believing that the lake concerned the life quality of people in Xiamen, she was very cautious when making decisions to address problems.

In our interview, Lin pointed out the importance of managing the lake: "Yundang Lake is the kidney and lung of Xiamen, and also a large drawing room for welcoming guests to Xiamen. "

※ Using a "three-in-one approach" for the integrated management of Yundang Lake

After reporting for duty, I formed a special bond with this lake.

In 1996, after graduation, I was assigned to work at the Division of Flower Production of Xiamen Municipal Government. At that time, Yundang Lake had completed the first two phases of integrated management and water quality and environmental conditions had been improved. There were still some problems like the pungent smell from the 11th and 12th south floodwater draining channels. As treatment was not thorough, sewage continuously flowed into the lake.

▼ People jogging by Yundang Lake in early evening (photo / Lin Yongchang)

Introduction to the speaker

Lin Xueping, deputy director of Yundang Lake Management Center, former deputy director of Flower Production Department of Xiamen. She joined Yundang Lake Management Center and has been engaged in the integrated management of Yundang for 10 years.

In 2008, the municipal government realized that this lake should be managed at the regional level and decided to merge Management Division of Yundang Lake, Management Division of South Lake Park, and the Division of Flower Production into the Management Center of Yundang Lake. So I was transferred to the center. Before 2008, my work was mainly related to landscaping. After 2008, my work refocused on the complex integrated management of Yundang Lake.

In the past, this lake was actually an open bay, called Yuandang Bay. In the 1970s, Xiamen started reclaiming land from seas by building dykes, and the lake was mainly used for aquaculture. The West Dike was also built. But years later, the environment of the lake began to deteriorate. Over a hundred enterprises and institutions congregated around the lake, and industrial and domestic sewage was discharged into it. The water thus became polluted and fish and shrimp disappeared. In the early 1990s, the government built the first sewage treatment plant here.

In the late 1990s, more and more people came to live and work around the lake, and the sewage began to exceed the plant's carrying capacity. As a result, the government built another plant. The lake not only blocks sewage, but also undertakes flood control and drainage. So in 1997, a pumping station was constructed to regulate its storage capacity, protect coastal environment, and improve water quality.

1
2
3

1. Since it was put into use in 2009, Yundang Acadamy has become a cultural platform for promoting traditional Chinese literature in the new era of "three-in-one", including Sinology education, cross-strait exchange of traditional culture, and special studies of traditional Chinese literature.
2. Predecessor of Yundang Academy
3. The current location of Yundang Academy, a combination of culture and gardens

In February 1991, the Management Division of Yundang Lake was established, and conducted its daily management. In 1997, the 10th Standing Committee of the Municipal People's Congress passed *The Management Measures for Yundang Lake in Xiamen*. Later on, management of all daily work related to the lake conformed with this regulation, including punishments for improper sewage and waste disposal; reasonable green space planning around the lake; prohibition one fencing areas of the lake for private use; no arbitrary land reclamation; and rules against most kinds of fishing except angling. These rules were clearly stipulated in the regulations. Therefore, the Urban Construction and Management Supervision Squadron was assigned to undertake urban construction management in the lake area.

In 2008, the Management Center of Yundang Lake was founded. In 2010, Songbai Park and Jiangtou Park, located at the upstream of the lake, were also incorporated into the management framework. In this way, the whole water system, upstream and downstream alike, was put under unified management.

So far, I have discussed how the management departments of the lake evolved. Now I'd like to go into further detail about how we managed the lake. Since the first day we initiated integrated management, we aimed to manage the lake and protect the western sea areas under the guiding principle of "intercepting sewage for treatment, dredging and building up shores, getting waters flow and improving the environment".

In 2005, we launched a large-scale sewage treatment and dredging in and around the lake. In 2009, we began sewage treatment in Songbai Lake and Tiandi Lake at the upper stream of the lake. We prioritized urgent tasks and started from the easy ones.

In 2011, we completed our basic targets for the integrated management of the lake. But why did the lake still smell? It is because the lake was a salty lagoon in an isolated coastal zone. Adjacent pollutants flowed into the lake with rainwater, and pollutants in the lake had no outlet to be discharged into the sea; and the lake lacked self-cleaning capacity. Even if sewage ran into the western sea areas, it would return with the tides.

Geographical position and functional characteristics of the lake necessitate continuous management efforts.

In recent years, with accelerating urban development, the lake management here has also continued apace.

Yundang Lake is in the heartland of Xiamen Island. Integrated management shows Xiamen's determination to rehabilitate local environment. But it takes time. So it is quite difficult to preserve an ideal environmental condition. We may often say it is normal when the lake smells and we can only alleviate pollution to some degree with feasible approaches.

Actually, pollution here is closely linked to how people use the water. If we all save water and have protection awareness, the lake will be exposed to fewer negative effects. Besides, waste water must run through the municipal pipe network rather than being randomly discharged. But few people have such determination or protection awareness. We have noticed that in communities surrounding the lake, residents would directly pour waste water from latrines into the lake. Such pollutants would deposit in the lake, and it takes long to clean them.

People have become more proactive in protecting this lake as time goes by. I believe when people complain about the foul smell, it shows they care. If they didn't care, it wouldn't matter to them. People live, rest and exercise around the lake, and of course they care about its environment. As a member of the Management Center, I share their feelings.

※ Getting the public involved in the management of Yundang Lake

While working on lake-related projects, I always believe that we are here to serve the lake, not to manage it. We aim to create a better environment for all. So in every decision we make, we welcome the public and experts from all fields to offer advice. The lake needs a long-term integrated management mechanism. Hasty decisions could jeopardize all previous efforts. As a public resource, the lake needs to be developed, utilized and well protected, and none of the three should be missing from the overall picture.

In our work, we often conduct surveys to learn about public opinion like the design of a footpath. We also set up the post of "Citizens first" to solicit public opinions and accordingly make due adjustments.

In face of problems and disputes, we need to take into account the demands of different parties.

For example, in recent years, it is common to see some people talk loud in the lake park. They disturb the others in the park. To deal with this,, we have talked to the citizen representatives. After discussions, we decide to put a noise monitoring device at the side of the road. When people walk by, they can see how much noise they are making and keep it down.

We use such flexible measures to accommodate different demands and make all happy. Besides, we stay in contact with the municipal government, the People's Congress and local communities to learn about their needs.

Current regulations allow for angling in specified areas, but other types of fishing are forbidden, because sluices in some areas are home to egrets. If anglers throw away their rods, it might affect the egrets. So lake management should also take the environment into consideration before taking any action.

We are very conscientious in our work. If the environment is damaged, there will be far-reaching effects. When people say that the water has become foul and smelly, it is actually very easy to make it clean again. Just throwing in some chemicals will do the trick. But as an environmentalist, I know that once we do that, it will take generations to clean heavy metal pollution in it. So we are very meticulous with the environment. We only take positive measures to protect the environment and the biodiversity and try to enhance the self-cleaning capability of the lake. For example, we attach great importance to enhancing the food chains of the ecological environment.

Apart from this, we are very careful when introducing plants. We all know that Yundang Lake has a marine environment and has 7 varieties of artificial mangroves, and they all grow very well. If new plants are introduced, no one can guess what they will bring to the lake.

Our management efforts are based on the concept of non-intervention. For example, we limit the number of visits to the central island of the lake to conserve it as a wild natural environment so that egrets can inhabit and reproduce on the island freely. We also conduct constant inspection and surveys over the island for analysis.

This is a kind of static protection.

Some suggested that egrets should be driven away as they always pee and poo on the railing. These kinds of suggestions will not be adopted. From the perspective of the environment, birds are a part of the ecosystem here. The more the biodiversity, the better the water can clean itself. In this way, the environment here can become increasingly natural, beautiful and healthy.

※ Cleaning lake waters will benefit the western sea areas in return

If Yundang Lake is properly managed, it will greatly help improve the environment of the western sea areas.

First, sewage interception and treatment will at least ease pressures on the western sea areas. But for our efforts since the 1980s, decades of pollutants and sewage would have drained into these areas and the ensuing pollution would be unimaginable.

Second, we have planted mangroves and carried out some other projects for ecological observation and conservation, which also greatly improves the environment of the lake. Improved air quality also benefits the overall marine environment. Actually, the lake is the lung at the center of Xiamen, providing a large amount of oxygen for humans to breathe. It plays a positive role in improving the air quality here and the whole marine environment.

Water in the Yundang Lake flows from the western sea areas, but during some months of the year, the lake water is even better than that of the west sea areas. In the past, if the western sea areas had red tides, it would be the same in the lake. But now, even if this situation occurred, the lake was immune. This shows that to some extent, compared with the western sea areas, the water quality of the lake has improved.

The most important function of the Lake is flood control and drainage. Without this function, Xiamen would be submerged during flood seasons. Although its water area only amounts to 1.6 square kilometers, the lake works to prevent floods over 37 square kilometers of urban area. Every year, before rainstorms come during typhoon season, gates to the lake are opened to intercept waste water on sunny days and drain it down into the lake through draining channels.

But this has rendered the scenic and flood prevention functions of the lake incompatible.

In management, the largest problem lies in balancing the scenic and flood prevention functions. For flood prevention, the gates of the lake must be opened to receive pollutants, while to improve the scenery of the area. Pollutants should not be allowed and gates of the lake must be closed to raise the water level and ensure beautiful scenery. For many people, the lake is a garden. But in flood seasons, the lake serves as a reservoir. Water around the region floods into the lake, and thus ensures the safety of the region.

After the flood season, when it turns cold and pollutants gradually drop, the lake becomes pleasing to the eye with its clear water. To resolve this issue, in the future, if we are able to build a large tank for flood prevention and sewage interception, all sewage from surrounding areas will be drained into the tank and directly treated in sewage treatment plants. Then, the lake will be free of pollution during the flood season and these two functions will be compatible.

But it does not mean that the new treatment tank will bring no pollution. With heavy rainfall from the south, even if sewage does not flow into the lake directly, it is bound to enter the western sea areas and then the lake through water circulation. So it is not possible to eliminate pollution. Environmental and resource issues will always exist in the area.

Later phases of management over the lake have followed the principle of "intercepting sewage for treatment, dredging and building up shores, getting waters flowing and improving the environment. " We are very proud that management efforts have received help from the whole city. At the beginning, we invited experts from all over the country to discuss how to manage the lake, and the principle they proposed has stood the test of time and remains practical today. Since the very beginning, we followed the "integrated management" principle. But some projects take time. It is not possible to completely intercept all sewage under certain circumstances, and we must be patient. A good example is the integrated management of Songbai Lake. Before 2010, the lake was extremely smelly, but with the construction of 23 gates for sewage interception, water has gradually become clear.

Some say that there are dead fish in the lake, but the reason is that they are fish needing to live in a marine environment. Once flood seasons come, when fresh water floods in and pollutants from the upstream pour in, some fish will die. But generally speaking, with sufficient diversity, they have strong resilience and are in good form without additional efforts. We are very cautious when managing the ecosystem for fear of disrupting food chains. Although humans manage the environment, we cannot disturb the ecological system to meet our own needs. We

▲ On July 30, 2008, cleaners were cleaning dead fish at Yundang Lake.

There is A daily scene of harmony between man and Nature in the Egret Island Park. (photo / Zhang Yongjie)

must protect the lake.

※ We shall continue to uphold the "sixteen-characters" management principle

Lake management here is coordinated and integrated, which is multi-faceted and is based on data rather than empty talk. For example, we have monitored its water quality since 1999, and will continue to do so.

Multiple generations have engaged in managing Yundang Lake. As early as the 1980s, the older generation has already gotten to work. Over decades, I believe most of our management efforts have been successful, and the management principle should be upheld. But currently, we need to make improvements to relevant regulations. *The Management Measures for Yundang Lake in Xiamen*, formulated in 1997, is not quite applicable today and new amendments must be made.

Laws and regulations are required to conduct environmental conservation and improvement. In implementation, they also need to be amended according to realities. For example, when *The Management Measures for Yundang Lake in Xiamen* was formulated, the lake was mainly used for flood prevention and there were no specific regulations over dog walking. But walking one's dog is quite common and the footpaths of the lake area are very narrow, so many problems have emerged due to this. As the lake has been transformed into a belt-shaped park through landscaping, constructing upgraded footpaths and incorporating nearby parks, it seems reasonable to forbid dogs from entering.

Additionally, when *The Management Measures for Yundang Lake in Xiamen* was formulated in 1997, Lianyue Road was incorporated into the lake region, but not the Songbai Lake and Tiandi Lake. But in reality, water from these two lakes also flows into Yundang Lake and they belong to one water system. But as this regulation does not include them in its framework, the regulation does not apply to them. In the future, we will make amendments to the regulation to include them.

I think that in the future we should integrate coastal management here with other bay areas in Xiamen, like the western sea areas, the eastern sea areas and Wuyuan Bay.

Compared with 20 years ago when I first arrived in Xiamen, the biggest change in Yundang Lake is that it has become increasingly beautiful. Especially after the footpath here was completed, it became very convenient. Some day-to-day management work of Yundang Lake is related to the Marine Management Center and other marine departments, as the lake is in nature a closed bay in a coastal zone. Before any of our activities that concern coastal management and construction, we will solicit opinions from coastal management departments. For example, as water quality of the lake is closely associated with that of the western sea areas, when we planned the construction of a new pumping station, we solicited opinions of these departments regarding how to make best use of the sea areas. Or when we explored the prospects of transferring water from east to west, we also solicited their advice.

Transferring water from east to west refers to water diversion from the eastern sea areas to revitalize waters in Yundang Lake. This program was initiated and proposed by a delegate of the Municipal People's Congress. The People's Congress and the municipal government attach great importance to the program and demanded that we consciously and scientifically analyze the pros and cons and feasibility. After over a year of study and analysis, the Shanghai Water Conservancy Exploration and Design Institute joined with us to draft an initial plan. There is

▲ Egrets perching and foraging

still much work.

Currently, there are many ongoing projects in Yundang Lake. One is the construction of an additional pumping station. It is estimated that after the station is completed, water exchange will increase from the current 1.3 million cubic meters per day to 1.9 million cubic meters. Currently, lake water is completely exchanged every 3 day. And after the station is finished, it will take only 2 days. It will not only bring additional vitality to the lake, but also enhance the lake's capabilities in flood prevention, as its rating will be elevated to prevent floods "that may be seen only once every 50 year". So it will help more in protect lives and property in Xiamen.

I hope that day will come soon.

Zheng Jinmu

The Economy of Xiamen Features the Marine Economy

▼

Zheng Jinmu said meaningfully, "the ocean is the largest treasury on earth with extensive resources. With the economic globalization in the 21st century, people are facing increasingly serious problems with the environment and resources. Thus, research, protection and development of the ocean have become the common concern for all mankind, and the marine economy is the field with the most strategic potential for years to come. " As an important Chinese coastal city, Xiamen should give full play to its strengths and actively participate in marine development so as to promote the development of the marine economy.

In the early 1980s, Zheng Jinmu served as the Vice Director of the Xiamen Planning Commission. In 1986 when the Xiamen Economic and Social Development Strategy Office was established to solve issues related to the strategic development of the Xiamen Special Economic Zone, Zheng Jinmu was made the Adjunct Director. Leading the Planning Commission for a long time, Zheng Jinmu has worked constantly to bring out new ideas, find solutions, write reports and solve problems. He is a hard-working explorer and active advisor who seriously studies the economy of the Special Economic Zone and continuously provides creative ideas. The Integrated Coastal Management Leading Group was established in 1993 and Zheng was hired as a member of its Expert Panel. From then to 2002, he has continuously put forward targeted and forward-looking views surrounding "reform" and "development" of the sea. "Sometimes, provincial ocean planners use Xiamen as a model and the provincial government often takes advice by experts from Xiamen."

Zheng Jinmu believes that a good memory is inferior to even a sodden ability to write. When he sees good policies or views, he writes them down. Zheng Jinmu said that he shuffles his notes when he does not have a definite train of thought and rejoices at his foresight. 80-year-old Zheng hardly slept the night before his interview. As a consultant, he took time to prepare for the municipal discussion about the orientation of Xiamen. The reporter borrowed the notebook and flipped to a page randomly. The notebook was densely filled with notes written by pencil and recorded the orientation of the city in detail. Zheng also highlighted one of paragraphs: "The orientation of a city includes the orientation of industry, function and nature, among which, the first is the basis, the second is the core and the third is the soul. Therefore, the orientation of nature is the most important since nature decides the functions which lead the development of industry and determine planning, construction and development of a city." At the end of the notes, he specifically excerpted the orientation of several areas such as Hainan, Shenzhen, Zhuhai, Shantou and Pudong.

Zheng Jinmu said, "The State Council approved Xiamen's positioning as a Special Economic Zone of China and an important coastal city, port and tourist destination in southeastern China. So, if we cannot protect her oceans, can the port and tourist city continue to exist?"

For 80-year-old Zheng Jinmu, being a member of the Expert Panel responsible for coordination of coastal management in Xiamen has been a proud and unforgettable experience. He often complains that his memory has deteriorated with the growing of age and that he often forgets where he puts things. But when we visited him, with the help of a ladder, he swiftly fetched out a red book stored for more than 20 years at the highest shelf of the bookcase. That was the offer from the Xiamen Marine Expert Panel, the Xiamen Integrated Coastal Management Coordination Leading Group!

※ Reporting on the ocean special program annually as a routine

A meeting reviewing the Outline of Xiamen Marine Economy Development Plan ("The Plan") was held on September 1st 1997. A panel of marine experts and leaders from municipal departments attended this meeting where Professor Hong Huasheng was elected as the director.

Introduction to the speaker

Zheng Jinmu, former director of Xiamen Planning Commission, was appointed as a member of Xiamen Marine Management Coordination Leading Group in 1993. From 1993 to 2002, focusing on "reform" and "development" of the ocean, he provided many precise and forward-looking insights.

During the meeting, the panel listened to the instructions delivered by director Xu Mo from the Xiamen Planning Commission on behalf of the editor group of the Plan and deliberated on the Plan from different perspectives and reached the following suggestions:

The Plan reasonably defined the scope of the Xiamen marine economy, and was realistic and forward-thinking. Closely following the strategic target of building a modern and international tourist port city, the Plan accurately reflected the overall approach and objective of marine economic development and orientation of every industry. It put forward the distinctive approach of function-based developments, and would guide the development of Xiamen's marine economy.

The followings are excerpts from a reviewing report from the Expert Panel of Xiamen Integrated Coastal Management Coordination Leading Group where I worked. "Xiamen pays much attention to the sea and marine economy and there are several points for reference: First is to establish a panel whose Director should be the Executive Vice-Mayor and experts should be real experts on the ocean. Second is to make plans and assessments on economic operations. It is difficult to manage the marine economy top-down, as it involves several different industries. Therefore, promoting development bottom-up in every industry through economic operations evaluation holds the key for success. It is fair to say that Xiamen has set a precedent in this regard in China. Third is to divide the sea areas according to its functions. Fourth is to introduce a compensatory mechanism when developing coastlines."

Working on economic issues for a long time, I participated in the whole planning of Xiamen. The Xiamen Marine Economy Development Plan is only a part of the tenth 5-year-plan. Besides the 5-year-plan, the Xiamen Planning Commission has several special programs which involve ocean planning. Before retirement, I served as the director of the Commission, which was responsible for both projects review and funding. If there is a project, we actively evaluate it and reply. Any plan involving coastal management, as long as reasonable, will receive funding from us.

As an important coastal city in China, Xiamen should actively explore marine resources and develop marine economy. Xiamen covers a small area and lacks land resources. The main resource of Xiamen is its sea. Its advantage lies in its ports, abundant marine resources and strong ocean technology. As an old saying goes, "living on what the land and sea can give. " Xiamen should fully explore its marine resources since it depends on the sea for the economic and social development. Xiamen should not only focus on the land covering an area of 1699.39

1. Xiamen International Ocean Week has become an important part of citizens' marine cultural life. The picture shows guests visiting yachts. (photo / Wang Huoyan)
2. People visiting the yacht show (photo / Yao Fan)

square kilometers, but also the water with an area of 340 square kilometers. The coordinated development of land and sea can help alleviate space pressure we face.

The key to the success of a city's economy is to find a path suited to its own conditions. When the economy has distinctive features, it becomes competitive and long-lasting. The greatest advantage of Xiamen is the ocean. Only by vigorously developing marine economy can Xiamen form its own features. In this regard, marine biotechnology, which is a high-tech industry with great potential, might be a good start. We could leverage our strength in marine science and technology as well as abundant marine biological resources to develop marine medicine and health products.

As stated above, Xiamen has limited land resources for development. Resources must be transported from other places, which restricts the development potential of Xiamen. The space for development on land is increasingly diminishing and the demand for resources is constantly

increasing, which requires us to resort of the sea in the future economic development. In order to promote sustainable economic development, we must transition from a land-based economy to a multi-level and large-scale development of both the land and sea. Therefore, every year's economic prospectus must specially include plans for marine development, so as to build the marine economy into a pillar industry of Xiamen.

For example, the Municipal Bureau of Ocean and Fisheries held a review meeting for Xiamen Ocean Economy Operation Assessment on December 13th, 2012. After deliberation, we agreed that the development of the marine economy would relieve our resource pressure and environmental bottlenecks, upgrade our industrial structure and strengthen our attraction and influence as an important economic center in the region west to the Taiwan Strait. "This assessment report is conducive to better understanding the operation of our marine economy, in particular the existing problems and development trends. By providing references for marine economic development, it can boost the building of a robust urban marine economy. " As is shown by the report, Xiamen has been equipped with an improved social environment, economic strength and infrastructure to develop the overall ocean economy.

※ Relative standards of the reviewing meeting

The ocean is the lifeline of Xiamen because it is Xiamen's main strength. So how does Xiamen go about developing its economy? At that time, the Xiamen Development and Reform Commission put forward a plan to develop an economy featuring Xiamen characteristics. The feature of Xiamen is its sea, and when Xiamen leverages its sea, it becomes competitive economically.

Some industries are directly linked to marine economy such as ship-making, fishing and transportation and some are not. Experts did a survey in the marine-related government departments to find out which area under their jurisdiction belonged to marine economy. Then the research group evaluated the operations of different industries related to marine economy. After putting together the whole picture, the expert panel then proposed a targeted development plan for the next year. Experts went to every relevant department involved to provide perspective, and the fruits were shared by all. Such a win-win mode of development is the invention of Xiamen. Another unprecedented practice by Xiamen is to introduce a compensatory mechanism for coastline development. People value resources when they are managed through markets. Xiamen Island covers a small area, but the panel strongly opposes further land reclamation, as the the used curvy shoreline of Xiamen Island now becomes smooth, making the Island look like a ball., which is the result of reclamation efforts. Therefore, control is necessary.

▲ The maintenance of marine environment requires the efforts and support of the public. Years of efforts to promote laws and regulations of marine management by Ocean & Fisheries Bureau of Xiamen have been widely recognized by the people and higher authorities.

I retired in 1996, but from then on, I made every effort to participate in marine planning and management as a reviewer. As reviewers, we are responsible for evaluating whether each project is adequately planned, the research target is specific, the framework is clear, the contents are comprehensive and the research methods are reasonable. We decide whether the project is feasible based on the sufficiency of relevant research, time and technical support. We need to make sure that thorny issues have been addressed in project plans. For example, we often look at whether the target is set too high or too low or whether the main industry to develop is clear and the spacial layout is scientific. The materials are often submitted for us to review two or three days before the meeting where reviewers speak out their opinion one by one and offer suggestions.

Different views are aired at the discussion group, but this usually does not happen in the reviewing group, because at this round, the submitted materials are relatively sufficient, feasible and targeted. Therefore, reviewers usually affirm others' suggestions and then offer some additional advice . We are scrupulous when voicing our opinions. The research group does not propose projects carelessly and our reviewing group takes our job seriously. We focus on what should be developed in Xiamen, whether it is in accordance with the Central Government's requirements, whether it conforms to the municipal planning standards, whether it is in line with the practical development of the sea, whether the planning is too radical or conservative, whether the direction is correct and whether the priority of development of an industry is appropriate.

On the14 April 2011, the Xiamen Ocean and Fisheries Administration held a review meeting over the proposed Development Plan of Building Blue Industry Belt and a Robust Marine Economy, and a reviewing group of 5 attended the meeting. After discussion based on the reviewing principles, the reviewing group concluded that while marine development had become a national strategy and priority for all coastal provinces and cities, Xiamen should take the lead in developing marine economy and secure first-mover advantage in the fierce economic competition. At the same time, Fujian province launched a pilot project to develop marine economy with the strategic goal of "building a blue industrial belt along the Taiwan Strait and becoming a strong marine economy"and required every prefecture to submit relevant plans and pilot proposals. For Xiamen, the compilation of Development Plan of Building Blue Industry Belt and a Robust Marine Economy and making pilot proposals are instrumental to grasp opportunities and accelerate the transformation of economic development. They serve an exemplary role for developing the economy west to the Taiwan Strait. Even today, the review remains targeted and forward-looking and reading it just brings the pleasant memories of marine development back.

The reasonable development and employment of marine resources is fundamental for human survival and sustainable social development. During Xiamen's Integrated Coastal Management, I have noticed that Xiamen enjoys an advantage in terms of marine ecology and marine resources and I thought that Xiamen's marine development has the following six pillars:

First, Xiamen should make use of its high quality ports to vigorously develop a shipping industry. With the Port of Kaohsiung as its target, Xiamen should develop an international transit business so as to make Xiamen port into an international container hub. Second, Xiamen should rely on its ports and bonded areas to promote a modern logistics industry. By vigorously developing shipping, distribution and comprehensive logistics, Xiamen can build itself into a center for logistics on the southeast coast and in the network of international logistics. Third, Xiamen should take advantage of its marine resources and technology to develop marine medical and health products. Fourth, Xiamen should fully tap the potential of its resources and develop coastal tourism. Only when marine resources are fully developed can Xiamen realize its potential as a center of tourism. Fifth, it should rely on its current industrial capacity and port advantage to vigorously promote port industries, such as the down-stream petrochemical industry, port power industry, shipbuilding industry, port mechanical industry, ocean environmental protection industry, industry of processing of ocean resources and export processing industry. Sixth, the ocean fishing industry should be transformed from output-oriented to quality-oriented. The priority should be placed on pelagic fishing, non-polluting cultivation and fishing for leisure.

※ Promoting the construction of a"bay city"

Some people wonder how Xiamen balances rapid economic development and sound coastal management. Despite economic progress, the marine environment is improving instead of deteriorating and the marine economy is booming with highlights. What are the reasons behind those successes? I think that building a "bay city" is an important initiative in this regard.

Xiamen put forward the idea of building a "bay city" before 2004. The Xiamen Special Economic Zone ranked fifth in China in terms of competitiveness after 20 years' economic development, and its numerous economic indexes ranked at the top among many cities. However, located on a small isolated island and with a small economic size, its development has been limited.Therefore, it has had to expand its development space by transforming from an island city to a bay city. Only in this way can Xiamen strengthen its comprehensive competitiveness and maintain sustainable economic development. After 20 years of construction, space for development has become increasingly limited on the island and the outer reaches of the island need to be explored. At the same time, the industrial structure on the island needs to be adjusted and the manufacturing industry needs to be transferred off island gradually so as to develop modern service industry on the island and build a free trade zone.

Xiamen is a city surrounded by several bays. Therefore, it enjoys good ports, a beautiful coastal landscape and abundant marine resources. It should build along the bay and take full advantage of ports to develop industries. Building a "bay city" also helps connect Xiamen with Zhangzhou and Quanzhou to create a closely interwoven economic zone in Southern Fujian province. Because the territory of Xiamen is quite small, it needs to rely on the whole region in its development.

The Minnan Delta along with the Yangtze River Delta and the Pearl River Delta are all important development areas designated by the Central Government. Xiamen, Zhangzhou and Quanzhou are neighbors and share a similar culture. Despite the similarities in economic structure, their development level and priorities differ. Because every city has its own features and economic advantages, they are complementary to each other, it would be better to break the administrative boundary and promote economic integration in the region. These areas should focus on their comparative advantage and cooperate in utilizing resources in the region so as to upgrade industrial structure and layout. The resultant economy of scale and efficiency in material consumption can boost overall returns and comprehensive strengths and lead to rapid development of all.

If a city wants to boost its economy, it must become a regional center with certain economic scale and strength. It would be impossible for Xiamen to become a regional center by development on the island. Therefore, the city must be expanded and the coordinated development with Zhangzhou and Quanzhou accelerated. On the basis of cooperation with Zhangzhou and Quanzhou, Xiamen should extend to the west by strengthening coordination with Longyan and Sanming, and to the north by strengthening coordination with Putian and Fuzhou, so as to build Xiamen into a real regional center. At the same time, it should seek a larger scale regional cooperation with southeastern coastal cities and provinces, thus becoming an important center on the southeastern coast of China.

How to build a bay city? At the time, the Expert Panel and the Leading Group held a discussion, where they determined the positioning of Xiamen as a tourist port city, a decision which received much attention

from the Xiamen Ocean and Fisheries Bureau. The Vice Mayor Pan Shijian, responsible for marine affairs, often invited us to visit the coastline to determine places to be returned to the sea and shorelines to be protected.

My thoughts about Xiamen building of a "bay city" are based on these field trips and the discussions by experts; To build a bay city, Xiamen could take the approach of improving its island, expanding the bay area, enlarging the hinterlands and highlight the two wings in cooperative development. The east and west wings of the island are the future direction for city expansion. Therefore, Xiamen should develop these two wings simultaneously with different strategies. While developing the west wing, the focus should be managing the western sea

▼ The old Haicang was a small, desolate fishing village. Xiamen, a city that's been chasing the tide of the time, has turned Haicang into a prosperous place. (photo / Wang Huoyan)

areas and the developing Maluan Bay. In the east wing, the priority is planning for important infrastructure projects and pave the way for further development of eastern areas so as to narrow the gap between the two wings and grasp potential opportunities.

In terms of the layout, the bay city was expected to cover Xiamen island, Gulangyu, Haicang, Xinglin, Jimei, Datong and Liuwudian by 2010. The idea at that time was that after the adjustment of functions, Xiamen Island would mainly develop high-tech industry, logistics, tourism, commerce and trade, finance, real estate, culture and education and gradually transform into a free trade zone, while Gulangyu Island focuses on tourism and culture, Haicang port logistics, petrochemicals and the export manufacturing industry, Xinglin mechanics, electronics and railway transportation, Jimei tourism education, research and incubators. Datong should light industry and food processing, Liuwudian port logistics industry and light processing, and Dachengdao trade to and tourism with Taiwan.

While building a bay city, it is important to adjust administrative division. While keeping the total number of administrative districts the same, Xiamen should make adjustments to its administrative division to properly distribute production as suited to its positioning as a bay city. Therefore, we suggested that Gulangyu district should be combined with Ximing district into one administrative district. The same should happen with Xinglin and Jimei district. Haicang should be remain its own district and Tong'an district should be divided into two. Such practices are favorable to economic development and enable every district to form an economic cluster with distinct feature.

Today, these plans have become a reality.

Development of All Forms Should Be Based on Protection

Guo Yunmou

▼

Nestled, like a flower, at the foot of green mountains, Xiamen overlooks the sea and rocky coasts. This city hosts millions of households and it is the final destination for hundreds of rivers before they flow into the sea. Woven into the fabric of Xiamen, the seas provide the fantastic scenery that supports a booming tourism industry. There are gulfs, beaches, islands, lakes, wetlands, hot springs and other natural aquatic features. The scenery, unique to southeastern Chinese cities, gives Xiamen its abundant charm. Blessed with warm weather and blossoming flowers, days in Xiamen begin by greeting the sea. Guo Yunmou is a local who grew up by the seas of Xiamen, and has a special bond with the sea. Since his career started at the Fujian Institute of Oceanology in 1981, he has dedicated three decades to the seas around Xiamen.

His responsibilities mainly entailed sampling surveys and collecting data, seemingly very standard oceanographic studies. But the report he produced on the seas of Xiamen served as the supporting theory and data for Integrated Coastal Management. His comprehensive survey of the central and northern part of the Taiwan Strait has filled a vacuum in the oceanographical study of this area.

Guo Yunmou believed that he became closer to to seas of Xiamen through his different projects and surveys. He lived a simple life for over a decade, spending his days between the office, the sea and the shore. He never tires of the seas, and has maintained strong affections for them throughout. He would go for walks along the shore and take photos of the sea he loved as he passed by. His descriptions of the sea were so detailed that it was as if he were detailing the heirlooms of his own family.

Having grown up by the seashore, Guo Yunmou loves the oceans dearly, and has committed to protecting Xiamen's seas whole-heartedly and with firm action.

※ The Basic Survey of Tong'an Bay still plays an important role today

Born in Tainan, Taiwan, I moved to Xiamen with my parents when I was four and then settled here. That is the starting point of my special affection for the seas in Xiamen. Since then, the seas have always been an inseparable part of both my personal and career life.

The sea was my best friend during childhood. I remember clearly that in summertime, I was crazy about hanging out on the beach around Baicheng and Gulangyu Island. I loved swimming so much that I even managed to swim across the Xiamen-Gulangyu Strait. I have so many fond memories of the ocean, and to this day, it gives me great pleasure to think back on them.

Because I spent so much of my childhood playing on the seashore, I have always been interested in nature's secrets. That is why I chose geology as college major. I was deployed to Xinjiang right after graduating from the Beijing College of Geosciences (now the China University of Geosciences) in 1960, and I stayed in Xinjiang for almost 20 years. My research subjects included snow-capped mountains, deserts (including the Gobi), and grasslands among others, and I visited all of them in person. I didn't come to the Fujian Institute of Oceanography until 1981, and since then my studies have focused on the seas surrounding Xiamen, essentially pertaining to their geological and geomorphological features.

When I came back to Xiamen in 1981, the seas I was once so familiar with became my research subject. I was very excited. I was appointed as Director-General of the Geology Department and Deputy Chief of the Fujian Institute of Oceanography. Due to lack of original

Introduction to the speaker

Guo Yunmou, former deputy director of Fujian Institute of Oceanology, served as secretary general of Expert Panel of Xiamen Integrated Coastal Management from 1995 to 2003. He has organized and participated in major research projects, including Renovation of Coastal Areas in Eastern Xiamen, Maluan Bay Causeway Opening, Causes behind Turbidity of Xiamen Waters, Issues related to Aquaculture in the Western Sea Areas, Adjustment of Zhongzhai Line along the Huandao Road and Protection and Utilization of Uninhabited Islands. He has put forward solutions and recommendations, most of which have been adopted by the municipal government.

research on the seas of Xiamen, data on the hydrological and geological features of the seas and gulfs was insufficient, and on certain study subjects, there was no data at all. So after I returned, my work mainly focused on data collection on the seas of Xiamen.

The Basic Survey of Tong'an Bay was the first project I took over when I began working at the Fujian Institute of Oceanography. At that time, the Institute was newly founded and was not large. It was located on the former site of the U.S. Embassy on Gulangyu Island. The west sea had been well developed by then, while Tong'an Bay, including Dongzui Harbor at the mouth of the West Creek and the East Creek of Tong'an, and Xun Rivers north of Xiamen, despite being the largest of its kind in the seas of Xiamen, was in a primitive state. After we took over this project, all colleagues from the Institute went out on field research together. We did surveys on everything that fell into the categories of marine biology, hydrology, chemistry and marine topography. I was mainly responsible for the research on topography and siltation and its causes in Tong'an Bay.

I remember that in 1982 we went to Tong'an Bay every month. I had to closely examine the seas and coasts and do sampling surveys to see if there was any erosion or siltation. My wife

was also working at the Institute. She was mainly responsible for collecting information on river basins and analyzing data on the sources of sediment and siltation at the Tong'an Water Conservancy Bureau. This project took us a year to complete and we worked together to produce the report Sources of Sediment and Siltation in Tong'an Bay. After doing sampling surveys of water bodies and sediment and collecting data, we found that after the construction of the causeway, siltation in Tong'an Bay increased by 30 cm, which changed the hydrological conditions of the entire area. In the past, the seabed of Tong'an Bay was mainly made up of sand, and it was an ideal habitat for lancelets, so lancelets flourished there, greatly supporting the local fishery. However, in the 1950s and 1960s, after the construction of the Gaoqi-

▲ The view of Kinmen, from Xiamen (photo / Wang Huoyan)

Jimei Causeway, the existing Xixiang and Yongjiang waterways in Xiamen were cut off, forming a semi-enclosed bay. Coupled with the extensive coastal land reclamation of the past two decades, the hydrodynamic conditions of Tong'an Bay changed and resulted in further siltation, which gradually destroyed the lancelets' habitat. The flow of sea water slowed and deposited the silt it carried there. After decades of accumulation, sand on the seabed of Tong'an Bay was replaced with silt and was no longer suitable for the survival of lancelets, and they consequently migrated to the open sea.

Because of the special features of Tong'an Bay and to create reference for managing the seas

of Xiamen, we started the research project and provided the Xiamen municipal government and institutes involved in marine development with the most basic information on the changes in the seabed of Tong'an Bay. To our surprise, the paper, Sources of Sediment and Siltation in Tong'an Bay has become one of the most cited papers on this subject by experts and scholars.

Against the clear blue sky, high-rise buildings have appeared on the horizon. A picture which is half ocean and half city is unfolding slowly in front of us. In the near future, a new Xiamen landmark—a romantic tourism route will also emerge on the coastline between the sea and tall buildings. This romantic route, which will soon be lined with mangroves, man-made beaches, jogging paths, hotels and many other scenic elements, is designed to be the tourist-friendly "Golden Coast" of Xiamen.

※ The Comprehensive Study of the Taiwan Strait has filled the vacuum in this area

Until the 1980s, there was no survey on oceanographical conditions in the Taiwan Strait. When this project was assigned to me, I was very excited and had high expectations. For one thing, I thought I could fill the vacuum in this area through my work. Additionally, although I was Taiwanese, at a time when the two sides of the Strait were in a state of isolation, it was impossible for me to go to Taiwan. My siblings were still in Taiwan and we had been unable to get in touch for two or three decades. I thought perhaps this project could help me get closer to Taiwan.

1/2. In December 2003, researchers from Ocean & Fisheries Bureau of Xiamen, Xiamen University and he Third Ocean Research Institute landed on the uninhabited island.
3. Baozhu island is full of cultural attractions and natural landscape. (photo / Liu Fan)

The Institute arranged for me to go to Tianjin to borrow a boat—the Yanping, a one kiloton ship, for research on the northern and central part of the Taiwan Strait. Like our work on Tong'an Bay, this project, which involved all relevant departments, required over a dozen colleagues on board together with scientific equipment and instruments. The Yanping sailed along the midline of the Strait, but as the seas got rough, many on board developed seasickness. I also felt seasick, but was in better shape than the others because of my long time experience working on ships. Despite the conditions, we began collecting water samples and measuring the depth and temperature of the water in an orderly manner. Suddenly, a Taiwanese plane flew over us. Because of the political tensions, we were afraid of possible conflicts. Luckily, the ship we borrowed was a civilian vessel, so we didn't have to worry as much as our colleagues from the Third Institute of Oceanography of the State Oceanic Administration. The plane eventually flew away after circling around for some time.

In this trip, we spent over a week at sea, and would do so again in future trips. This project took us a year to finish. Because of seasonal changes of ocean currents, flow velocity, temperature and marine creatures, colleagues from the departments of hydrology, chemistry, and biology needed to accompany us on a trip to the Strait every month. But I only participated five or six times because the topographical features and sediments we geologists studied did not change seasonally. Our ship was once so close to Taiwan Island that I could see northern and central Taiwan clearly with my naked eyes. I really wanted to land on the island and take a look.
It took us more than a year to complete the paper, A Comprehensive Survey of Northern and Central Taiwan Strait, which provided basic research data on the Taiwan Strait. This paper made clear where there was siltation, which parts of the seabed were made up of sand and where currents were rising. It also provided useful information on fishing resources.

Through these projects, we became more familiar with the seas of Xiamen, and I felt emotionally closer to them. It became a routine for me to take a walk by the seaside and visit the islands. I would go out for every study and survey that I did, so I became familiar with every corner of the seas.

I visited all of Xiamen's uninhabited islands, large and small. Although conservation and development of uninhabited islands was not my responsibility, I found it interesting to visit these islands because I was curious about every aspect of Xiamen's seas. Huoshaoyu Island was terrific. On Huoshaoyu Island, there was a lake which attracted many egrets, water wells that provided fresh water and an area of sedimentary rock that looked like a blazing fire, earning the island its name, Huoshao, "burning in flame". Crocodile Island, which also had wells and a mangrove forest, was home to many egrets. The Chicken Island, at the mouth of the Jiulong River close to Haimen Island, had luxuriant vegetation.

In fact, I visited those islands so frequently that I couldn't remember how many times I had been there. Later when I was working as Secretary-General for the Marine Expert Panel of Xiamen Integrated Coastal Management, I often arranged trips for experts to walk around those islands and take a look.

Based on my research, around the year 2000, I submitted a proposal for tourism around the islands to make the best of their rich flora and fauna, and my proposal won an Award for Excellence.

※ The profiling of Xiamen's marine coastal zone provides the basis for integrated coastal zone management

Although most of my work at the Fujian Institute of Oceanography was about basic data collection and surveys of the seas, the results have become an important source of data for integrated coastal management. In 1994, after the establishment of the Marine Expert Panel, I was appointed Secretary-General, while retaining my position at the Fujian Institute of Oceanography. The Expert Panel worked not only on doing research and collecting datan, like the Institute, but also on proposing solutions. We had meetings on either utilization of ocean resources or marine environmental impact assessments almost every week.

In 1993, the Global Environment Facility, the United Nations Development Program and the International Maritime Organization proposed the Xiamen Demonstration Project on Marine Pollution Prevention and Management for the East Asian Seas. At the time, it could be counted as the largest and most comprehensive coastal management project in China. In February 1994, the Xiamen municipal government established the Executive Committee, the Office, and the Expert Panel for the Xiamen Demonstration Project. Scholars, officials and legal and academic experts from many institutions including the Third Institute of Oceanography of the State Oceanic Administration, Xiamen University, the Fujian Institute of Oceanology, the Fujian Provincial Institute of Fisheries, the Xiamen Oceans and Fisheries Bureau, and others, all took part in this Demonstration Project.

Establishing Environmental Profile of Xiamen's Coastal Zone was the first task of the Demonstration Project, and I presided over its compilation. Based on existing data, the research analyzed the health of the marine environment and the organizational structure, for example the specific functions and operations of agencies in charge of marine affairs. Additionally, it reviewed existing environmental laws and regulations as well as their enforcement mechanisms, and reviewed the sources of funding for marine protection. In short, Environmental Profile of Xiamen's Coastal Zone discussed problems related to marine

resources and environmental management in Xiamen, their causes, their order of significance, and potential consequences. The Profile has served as the scientific basis for subsequent management plans and as a model for sub-projects.

I was responsible for writing the first chapter—The Relationship Between the Environment and Economic Development of Xiamen's Coastal Zone. Data collected since the 1980s served as a basis for the study, but we also tried our best to gather the latest socio-economic data to the extent possible. Over the past two decades, many institutes and universities in Xiamen have studied the natural environment of Xiamen's coastal zones. In the past 10 years, government agencies have been responsible for environmental protection and coastal management have initiated a number of research programs. Environmental Profile of Xiamen's Coastal Zone was based on these efforts and concisely summarized the characteristics of the environment, marine resources and management structure of Xiamen's coastal zone.

1 | 2

1. After rehabilitation, Huoyu Island became a pioneer in the construction of the "Island Tour" project. (photo / Li Wanhan)
2. Prior to rehabilitation, the status of aquafarming on the outside of Firewood Island.

Yet I also met with some thorny issues during preparations, such as the lack of existing data related to environmental protection. There was a wealth of materials on marine hydrology and marine life, but not so much available related to integrated management, control and prevention of marine pollution. We lacked data on the sources of marine pollution, had no long-term systematic monitoring and analysis of changes in marine ecology and environment, and lacked a marine information system that involved multiple sectors and industries. We therefore often organized visits and seminars to update our database in order to tackle these problems.

It took us a year to complete Environmental Profile of Xiamen's Coastal Zone, a project which laid the foundation for the establishment of the Oceans and Fisheries Bureau of Xiamen. In addition, the five-year Xiamen Demonstration Project of Marine Pollution Prevention and Management in the East Asian Seas had yielded fruitful results, including mechanisms for integrated coastal management and prompt action by relevant authorities. We had accumulated

▼ Come to the seaside of Xiamen in summer. A hug with the sea is lovely and surprising. (photo / Gao Zhengquan)

a wealth of practical experience in just five years, formulating a comprehensive coastal management mechanism, a regional legal framework, an integrated monitoring network and training mechanism. These efforts also provided practical technical services for the integrated coastal management in Xiamen and provided a model in East Asia.

From 1994 to 1998, a total of 23 sub-projects involving over 200 colleagues were launched under the framework of the Xiamen Demonstration Project of Marine Pollution Prevention and Management in the East Asian Seas. Together, we successively finished conducting field investigation, data analysis, report compilation and promoted the results and relevant policy implications.

※ Managing Xiamen's seas has been my lifelong mission

By 1995, I had already turned 60 and was ready to retire. However, due to the Xiamen Demonstration Project, supported by the United Nations Development Program, the Global Development Funds and the International Maritime Organization, I kept working until 2003. Even after I retired, I continued to work for the Marine Expert Panel for a long time.

In the past three to four decades, I have seen Xiamen's seas become more and more blue year by year. The city has undergone profound changes as well. Prior to the implementation of the Integrated Coastal Management and the Marine Development and Management Ordinance in the 1990s, the seas of Xiamen underwent chaotic development and were overwhelmed by excessive aquaculture. Fishing nets were scattered everywhere, except for waterways.

After Integrated Coastal Management was put in practice, geological and hydrological conditions in Xiamen's seas have improved due to the implementation of marine functional zoning, the opening of Causeways, and newly developed beaches, and because of restrictions on aquaculture. Today's eastern romantic coastline used to be a barren strip of land, badly eroded and covered in sand. Now the beach is in a stable condition, and beachfront properties fetch high prices. The development of the eastern beaches was also a product of the Expert Pane's discussions and tests. Since ocean currents in the east flew from south to north, we built a dam along the northern coastline.

As a member of the Expert Panel, I mainly organized and participated in such projects as the Renovation of Coastal Areas in Eastern Xiamen, Maluan Bay Causeway Opening, Causes Behind Turbidity of Xiamen Waters, Issues related to Aquaculture in the Western Sea Areas, Adjustment of Zhongzhai Line along the Huandao Road and Protection and Utilization of Uninhabited Islands. I put forward corresponding countermeasures and proposals, most of which were adopted by the Xiamen municipal government. Research in written forms amounted to more than one million Chinese characters, covering marine science and technology, economics, environmental protection, functional zoning, among others.

The seas and oceans are my research subjects, yet they are also my lifelong friends. In my spare time, I have always enjoyed taking walks by the sea. On every walk, I make sure to take photos of the seas and coastlines. After some time, I would return to see if there have been any changes, such as siltation or erosion. Even when I am on vacation with my family, whenever I see the sea, I cannot help but take the camera and go back into work mode. In 1999 when we first went to Taiwan, we went to visit the east coastline of Pingtung, Hualien. We were amazed by the water there, which was extravagantly blue and clean, probably one of the best coastlines I have ever seen in my life.

Over the past 30 years, Integrated Coastal Management has achieved remarkable results in Xiamen and the seas of Xiamen have been well developed. Now the program is well known both at home and abroad. I think protection of our seas and oceans is vital because they enable the production and sustainment of our modern lives. In my opinion, the development and use of seas and oceans should be based on protection. Only in this way can the seas of Xiamen continue to prosper in a sustainable way.

Promoting the Sustainable Development of the Marine Economy

Peng Benrong

▼

The profile of Mr. Peng Benrong clearly shows that for most of his life, from a teacher at a secondary school to a civil servant, his work had little to do with the ocean. Since that time, however, he has set on a path to study marine economics and become an expert of the ocean. This is impossible to understand without learning about his story with Yundang Lake.

Mr. Peng said: "When I was a child, there were all kinds of water plants, but now there are none at all. " Growing up by the seaside, Mr. Peng loves the seas of Xiamen very much. Seeing the seas in Xiamen deteriorating, he was very anxious but didn't know what to do until he participated in the management of Yundang Lake in Xiamen. "This campaign is the first Integrated Costal Management (ICM) project that I have engaged in. Since then, I have become very interested in studying the marine economics and marine conservation, and have devoted myself to this cause wholeheartedly. "

In 1997 when Mr. Peng worked at the Investment Promotion Division of the Foreign Direct Investment Bureau in Xiamen, he was recruited to the Yundang Lake Management Working Group. He began to work with no clues at all and later started making contributions to the planning with his economic knowledge.

1. The busy Xiamen Port in the early 1930s
2. Today the Xiamen Port is still busy. (Photo / Hong Weini)

Introduction to the speaker

Peng Benrong, Professor of School of Environment and Ecology, Xiamen University, became a member of Expert Panel of Xiamen Integrated Coastal Management in 2003 and participated in the efforts to revise fee collection standards for the use of sea areas in Xiamen and to analyze the social and economic benefits of comprehensive rehabilitation of Xiamen Western Sea Areas.

Having worked in trade promotion for years, Mr. Peng suddenly realized how much economics could be applied to coastal management and how he could play his part in improving the seas around Xiamen.

After the Yundang Lake project ended, Mr. Peng made his decision. Instead of returning to his original post, he chose to study marine environment at Xiamen University. At the same time, he became a member of the Expert Panel for the Integrated Coastal Management Committee in Xiamen. With years of study and practice, Mr. Peng knows that we must follow rules as we develop the marine economy. Sustainable development of marine economy can only be realized on the basis of environmental protection.

※ Coastal management should give consideration to equitability and the interests of fishermen

In 1997 when I worked at the Investment Promotion Division of the Foreign Direct Investment Bureau in Xiamen, I participated in Xiamen ICM projects.

In 1997, Xiamen initiated a campaign for Yundang Lake management, part of which was to analyze the social and economic benefits of our efforts. Due to shortage of personnel who studied economics at the Management Division of Yundang Lake, they came to me for help. After I got on board, my main responsibility was to systematically assess the social, economic and environmental effects. When the project was completed, I did not return to my former job; instead, I decided to work on projects related to Xiamen ICM.

To discuss the necessity of implementing integrated management, and as required by my leaders, I provided my arguments from an economic perspective. For example, back then, many sea areas were used for aquaculture, and most ships that sailed in would be tangled in fishing nets or could not sail in at all. To avoid such conflicts in resource utilization, I made a monetary evaluation from a socioeconomic perspective, which concluded that 5 years of ICM could bring 2 billion RMB in economic benefits for this area. This result clearly pointed to the advantages of integrated management.

Xiamen ICM began in the western sea areas. Earlier, serious sediment accumulation necessitated expensive dredging every year. Against this backdrop, it was decided to implement integrated management in Xiamen's western sea areas. This involved several key projects. The first was to block water from entering these sea areas. In order to prevent terrestrial water from polluting the seas, we took a three-pronged approach to confine polluted terrestrial water to land. Meanwhile, we built many sewage treatment plants and focused on treating industrial sewage directly discharged into seas. Second, creating an opening in the Maluan Bay Causeway. Originally, Maluan Bay stretched dozens of kilometers. However, after reclamation, it measured only 3 km^2. As a result, its flow rate was very poor and a river mouth needed to be opened up to let sea water in. The third concerned the management of mud flats. Some mud flats required the planting of mangrove forests while others needed to be generally rehabilitated. Another difficult task was to persuade the fishermen out of aquaculture.

Back then, in the western sea areas, there were severe conflicts of interest between the shipping industry and aquaculture. According to our calculations, in this region, the shipping industry, or even tourism industry, could yield more profits than aquaculture. So it became necessary to make a plan to compel the aquaculture industry to leave the western sea areas.

As a result, we were faced with the issue of compensation. I proposed that we should provide fishermen with something more than compensation, because compensation was only a one-time fix, and without aquaculture, fishermen would lose their main source of income. Later, I suggested to the Management Committee that we should first provide free training to fishermen to enhance their job-hunting capability. This arrangement worked for those below 40 years old. What about those above40? After discussion, we believed the government should provide jobs to those fishermen, jobs like sanitation workers and security guards. Another solution that I proposed was to initiate a project to build houses for fishermen that they could then use as they saw fit, whether by renting it out, opening a restaurant or so on.

In my opinion, from initiation to adoption and implementation, the ICM of the western sea areas was quite advanced. Because the project was not only about restoration and management of the environment, it also took the interests of fishermen into consideration and emphasized social justice.

After initiating ICM for the western sea areas, we moved on to the integrated management of Xinglin Bay and the eastern sea areas. To sum it up, every management program should adopt a holistic approach integrating both land and coastal management, because there are many activities on land that are deeply associated with the seas. For example, in coastal land reclamation, though no harm is done to seas directly, when waste water from farmland reaches the coast, mangrove forests begin to wither immediately, disrupting tidal flow. Therefore, no matter how hard you try to protect the seas, marine ecology is bound to be affected by human activities on land. That is the very reason why we should adopt an approach integrating both land and coastal management.

※ The payment for sea use has become a new standard of reference for the nation

When I worked on Integrated Coastal Management in Xiamen, I helped to assess the important "benchmark price" for using sea areas. In 1993, the State Oceanic Administration adopted a regulation which set the benchmark price of sea areas per mu at 200 RMB. But back then the price for one mu on land was about 20,000 RMB. The price difference was too large. With the implementation of the strict national "red line" restrictions on land use and the subsequent increase of demolition costs on land, reclamation from the sea, which boasted low costs, became an important avenue allowing coastal provinces and cities to expand space. Like other coastal cities, to meet the need for urban development, Xiamen also set its eyes on the oceans and began reclaiming land from seas. In history, Xiamen had many winding curves of its coastline and thus was known as a "bay city". But sea reclamation resulted in the disappearance of many of these bays. So, some jokingly began calling Xiamen Island "a big round cake" or an "oceanic trench city". A typical example is how Yundang Lake has changed. Originally, it was a bay with a total area of $36km^2$. The scene of "Lights of fishing boats on Yundang Lake"

was one of the most famous and beautiful scenes in the local region. But after coastal land reclamation, it shrinks into a small lake of less than 2km^2.

After recognizing the problem, Xiamen decided to make changes. First, the benchmark price of using sea areas was raised, i.e., the price for sea area utilization. Back then, Xiamen was the first in the country to adopt a standard fee for sea area utilization, and we are honored that the first such national standard also took reference from our program.

Actually, a long time ago, Xiamen attempted to raise sea area utilization fee through market regulation. Back then, the new standard had not yet come into effect and Xiamen was among the first in the whole country to work on the issue. One of the programs that impressed me most was the auction of sea areas. There was a project called the Xiangshan International Yacht Club, an important development in Xiamen at the time. This project occupied a sea area as large as 700,000m^2, while the total sea areas available for bidding was 220,000m^2, all

▲ The locals of Xiamen say that looking from the west bank, Haishu East Island is beautiful, and there is endless tasty seafood. The locals of Dongyu say that, looking from the Dongyu Bay, Xiamen is a dream of prosperity and wealth. Today, Haicang Bay, which is the former Dongyu Bay, is realizing its urban dream. (photo / You Zefang)

used for sea reclamation. Initial estimation showed the fees collected for sea area utilization in the area amounted only to 10 million RMB. Some even suggested the area should be credited to investors at a low price so as to support this project. After consultations among several sides, the Municipal Government of Xiamen decided to determine the use of the sea areas by listing and auction. On the day of the auction, after over 20 rounds of bids, to the surprise of all in attendance, the final price reached 323.9 million RMB. After deducting 70 million of compensation costs, preparation fees and charges of supporting infrastructure, the real transaction price for the sea area was 1,154 RMB per m^2, 18 times the standard price. It is fair to say that this project was an auspicious testament to Xiamen's ability to allocate marine resources and improve efficiency and environmental protection of mud flats through market.

After that, we estimated the sea area utilization fees for Siming District and Huli District. I remember that according to the initial standard, the price was 180 RMB per m^2. Counting in the costs of reclamation and taxes, the price sat at about 1,600 RMB per m^2. But if you checked property prices, you would find the value to be as high as 20,000 RMB per m^2. or even higher, especially for seafront property. Such potential for profit provided strong motivation for sea reclamation. It also shows how low our old standard was compared to the real value of sea areas.

So I made clear two points in a subsequent discussion. First, sea areas are a state asset and belong to us all. If fees collected for their use are too low, the value of our sea areas will flee to the pockets of real estate developers. Second, a low standard will spur everyone into sea reclamation and thus cause much damage to the marine environment. My advice was taken and we established a new standard of sea areas fee, a very important achievement. Through thorough estimation, the price I calculated was not 180 RMB. If the price of commercial land were 6,000 RMB per m^2, the benchmark price of sea areas should be about the same.

In formulating the standards, I did not give a fixed number. Instead, the charges shall be collected on the basis of a certain proportion of the price for a similar area of land in similar locations for similar uses. That is to say, "different purposes, different charges. " For instance, if one area of sea is reclaimed for building houses, we need to know how much the developers intend to sell the houses for and to what extent seas contribute to those housing prices. If the land price is 100 RMB, with cost of reclamation, management, tax, and reasonable profits deducted, the rest is contribution from sea areas. So the final standard I make includes all the profits put together, apart from reasonable profits for developers.

Additionally, I also suggested collecting fees from developers as compensation for damaging marine ecology. Marine space not only acts as a factor of production. It is also home to the marine ecological system and provides various services to human beings. Though sea reclamation expands available space, it also damages marine and coastal ecosystems. Actually, ecological damage is a more serious problem than the overall reduction in sea area, as once sea areas are reclaimed, the services provided by the local marine and coastal ecosystem are irreversibly lost. Therefore, developers must not only pay for the value of marine space as a factor of production, i.e. the charges of sea area utilization, but also compensate for the value lost in the marine ecosystem when using the land.

In retrospect, the final fees for the use of sea areas of the Xiangshan International Yacht Club were far higher than the standard of the time, but these fees excluded compensation for damaging the natural ecosystem. To our delight, later on, Xiamen was the first to complete its legislation for the regulation of fees for sea area utilization. In May, 2010, the People's

Congress of Xiamen Municipal Government introduced a local law called *Regulations on Conservation of Marine Environment in Xiamen*, which incorporated marine ecological "compensation fees" as an important element. Currently, we are working on standards for marine ecological compensation fees and relevant implementation plans for the Ocean and Fisheries Bureau of Xiamen. I am convinced that after relevant details for implementing the compensation standards are settled, its introduction will play a positive role in protecting coastal wetlands in Xiamen.

※ Every square meter of sea areas reclaimed must be justified

In setting new standards of sea area utilization fees, I also recommended controlling the total sea area available for reclamation through ICM. Although it is not possible to stop the trend of sea reclamation, it is also not possible to let it continue unabated. Some think that the ultimate target of sea reclamation is to realize maximization of net social profits. On the surface, it seems obvious that the more sea is reclaimed, the more can be made in profits. But this is wrong, because damage to the environment has not been counted in. Sea reclamation brings profits, but also impacts the environment and we need to balance the pros and cons. Sea reclamation addresses the shortage of space in coastal areas, while it seriously damages marine and coastal ecological systems which people rely on for living and development. In a port in Amsterdam in the Netherlands, where I visited during a field study, they mentioned that it takes over a dozen of years to justify sea reclamation for just one km^2.

However, in some regions in China, a large area can be reclaimed in one stroke, but after a decade or two, people may regret their decision, though at the current stage of development, sea reclamation seems to be a very good method to utilize marine resources for economic development. So a lot of truths reveal themselves during the course of development. Now, what we need to do is to try our best to construct a model to analyze the maximum areas possible for sea reclamation, taking all profits and costs into consideration, including environmental and ecological factors after systematic calculations. Hopefully, this model can serve as a good reference for others to avoid future twists and turns.

In the integrated management of the eastern coastal areas, we encountered these twists and turns during the reclamation of sea areas for roads. In the first version of the project plan, the drafted area for reclamation was 30 km^2. Our Expert Panel believed this area to be too large. Then we went to the head of the Coordination Group and asked to estimate how much we

▲ On November 3, 2016, Asia's largest international luxury cruise ship, the 170,000-ton Royal Caribbean "the Carol of Sea", docked at Xiamen International Cruise Homeport. (photo / Zhang Qihui)

could bear to reclaim to safeguard the health of marine ecological system. For this, I especially studied a type of software for designing scientific models, and after a while we finally decided which areas and how much land was appropriate for reclamation.

At first, the figures estimated by the Expert Panel were not accepted by the Integrated Management Committee for the eastern sea areas, and after negotiation and consultation, we finally settled on the figure of 10km^2. At the same time, reclamation required dredging in other areas of the eastern coast surrounding regions to expand marine space, so as to make up for the 10km^2 reclaimed. Though this figure was still larger than the initial estimate of the expert group, it represented substantial progress and an improvement over the original plan.

I have always hoped to incorporate the formula for calculation of reasonable areas for sea reclamation into the legal framework, so that when faced with a strong motive for sea reclamation in the future, we will have legislation to ensure at least a minimum sea area is preserved to protect relevant bays. In this way, even if one day the Expert Panel were dismissed, the government would still have a scientific approach to determine areas suitable for sea reclamation in their project plans. It could also help guard against changes in administration and attitudes towards reclamation. In this regard, the management over Yundang Lake played an exemplary role. Article 7 of *Management Measures for Yundang Lake in Xiamen*, which came into effect in 1997, prohibited any organization or individual from reclamation in the lake. This ensures at least 1.5 km^2 of the lake area is available for holding flood water. This article means any organization or individual must amend the regulation before they reclaim the lake for land, which has protected the lake region for over a decade. Besides, the project called Xiamen Semi-closed Bay (Western Sea Areas, Tong'an Bay) Aggregate Sea Area Control Study, jointly conducted by School of Ocean and Environment at Xiamen University and the Fujian Institute of Oceanography and completed in 2006, demonstrated that the minimum acceptable area for sea areas in the western sea areas and Tong'an Bay was 51 and 83 km^2 respectively, thus only allowing 1.5 km^2 and 9.5 km^2 for reclamation in these regions. This result also provided a basis for the law-based control of total marine area Xiamen. I have learned that many experts and delegates to the People's Congress are promoting this legislation.

※ Both natural and man-made restoration of coast lines do good

There is another important aspect to the Integrated Coastal mManagement in Xiamen: the restoration of natural coastlines. Our target is to keep no less than 35% of natural coast lines conserved, a very ambitious goal.

What advantages do natural coastlines bring? Why should we take great efforts to restore them? To answer these questions, we must take a holistic approach to economic thinking and estimation. The marine ecosystem actively exchanges materials and energy with land ecosystems. That is to say whatever comes from the land, water, sand or other things brings energy to the oceans. If we artificially separate oceans from the land, and reduce the input of energy required by oceans, eventually, the whole marine ecosystem will be destroyed. With natural coastlines, it is easy to restore the circulation between oceans and the land.

Apart from natural coastlines, Xiamen has also laid many man-made sand beaches. Why should we build man-made sand beaches rather than leave the area to restore itself naturally? There are several reasons. First, it depends on whether the conditions are suitable for natural

restoration. Second, even if they are suitable, it also depends on whether it is the best option.

Maintenance of natural sand beaches is closely related to the marine environment. It requires a cyclical process. As tides rise and fall, sand is washed into the sea from beaches and then returned as the tides rush ashore. With erosion due to sea water dynamics, natural sand beaches are gradually lost. So ocean dynamics is the key. There are many restored sand beaches that became damaged years after restoration. The word "natural" does not always mean "good", and there might be other favorable alternatives.

Some say man-made sand beaches are just decorative, not a form of ecological restoration. Strictly speaking, it is not "restoration", because restoration implies the restoration of ecological

1. Xiamen Sea Recreation & Fishing Base is located on the outskirts of the Moon Bay on Xiaodeng Island.
(photo / Wang Huoyan)
2. The Underwater World of Xiamen is located in Huangjiadu, the east coast of Gulangyu Island.
(photo / Lin Shize)

systems to a natural state before they were damaged by humans. But in some places, there is simply no way to restore an area to its original state except for by human intervention. Of course, there may be some changes during the process, but all changes demand reasonable evaluation. In fact, this is a very important part of our work on the integrated management of coast lines.

※ Xiamen Model—making improvements alongside continuous expansion

Looking back on the implementation of ICM in Xiamen, I think on the whole we have done a good job. Though we had ups and downs, we made improvements with every new initiative. I believe Xiamen has already created a system of its own which deserves to be promoted as a model for several reasons.

First, before implementing ICM, it is very important to build a sound coordinating mechanism. The Integrated Coastal Management Committee as in Xiamen is headed by the Executive Vice Mayor, which ensures that its work is properly coordinated among departments. At the same time, the Marine Management Office of the Municipal Government of Xiamen was set up with heads of all government organs as members. In this way, whenever a problem occurred or conflicts arose among departments, especially those related to resource utilization, they could be coordinated and resolved, and thus overall coordination is enhanced. Second, a Marine Expert Panel was especially established to provide technical support during the process. In accordance with the experts"suggestions, a series of plans were made, including the establishment of Marine Functional Zoning, development plans for the marine economy, plans on proper utilization of marine areas, conservation and utilization of uninhibited islands, which all laid a solid basis for the scientific and regulated management of Xiamen's seas. Besides, we also adopted several regulations on Integrated Coastal Management, improved joint law enforcement mechanisms and forged synergy in this regard.

Second, management efforts over sea areas in Xiamen keep expanding and such expansion is very important because it shows vitality. Without an expansion of scope, after some time, when we have addressed all problems in front of us, the management efforts end. This is not correct, as in real life, problems cannot all be resolved in a certain period of time. From as early as the management of Yundang Lake to integrated management over the entire sea areas, we have continued upgrading our efforts, from simple management to ecological restoration. Xiamen did well in this regard, engaging in complex efforts such as opening the causeways to keep the marine ecology healthy and dynamic, restoration of islands, planting mangrove forests and laying man-made sand beaches.

All the examples above are expansions related to scope of work, but we also expanded in function by applying ICM to marine disaster prevention and reduction, like increasing protection from storm surges and management of industrial waste. These kinds of emergencies cannot possibly be tackled by one single department, but require joint and coordinated efforts between multiple departments.

As we expand the ICM, I support applying it to the development of a whole "blue marine economy". I proposed the concept of a blue economy. Blue represents health and a "blue economy" stands for a sustainable marine economy. In Fujian province, Xiamen is the best-positioned in that respect. Before we start, we must consider three elements. First, a blue foundation. Environmental resources and the ecological system are the foundation of the well-being of Xiamen citizens and must be well safeguarded. Second, economic growth. If management over marine ecology cannot translate into cash in people's pockets, they will lose interest. Third, social justice. In formulating plans for a marine economy, we must consider its inclusiveness and make sure everybody, including disadvantaged groups, can enjoy the benefits of a developing marine economy.

In summary, initiated to simply resolve the conflict over resources, the Integrated Coastal Management in Xiamen has eventually expanded into an integrated system for the sustainable development of the marine economy as a whole.

Currently, China is pushing forward development of the marine economy. Actually, our understanding of marine economics is also constantly improving. The earliest focus of the marine economy in Xiamen was the fishing industry, but now we also have a mature shipping industry and tourism industry. At the same time, we should bear in mind that while it is good to develop the marine economy, we also have minimum requirements for environmental conservation and ecological protection.

What direction will ICM take in the future in Xiamen? I think we need to strengthen coordination between the management of land and seas. Now that we have a bigger plan, we need to dig deeper in the details. We need to make conservation of the marine environment a main focus, and use it to direct people's activities on land and make better plans and arrangements for the future.

China has created many new concepts related to coastal management. Early on, there was the concept called Integrated Coastal Management, and now we have a new term called Management Based on the Ecological System.

In early times, we always considered what humans needed first in order to meet our short-term demands. So all our plans were intended to maximize profits in the short term. In contrast, now we propose management based on ecosystem, which aims to maximize our profits in the long run. By profits, we cover not only economic profits, but also non-economic indicators. For example, there are mangrove forests. It is difficult to imagine significant economic profits emerging as a result of their existence. But a mangrove forest can be a shelter for fish and keep the marine ecosystem stable. Meanwhile, wherever a mangrove forest is located, it can help stabilize the flow of water and reduce erosion. These merits are not directly reflected in the market, but they are important dividends indeed.

Riding the Wind and Waves and Piloting Coastal Management —Experiences from Integrated Coastal Management in Xiamen

If one song were picked to represent Xiamen, Waves of Gulangyu would undoubtedly be many people's first choice, because the song eulogizes Gulangyu and the ocean—two well-known symbols of Xiamen.

In the early morning of July 9, 2017, Beijing time, in Krakow, Portland, Waves of Gulangyu was played after Gulangyu was listed as a World Heritage site. The video of the ceremony was transmitted live across thousands of miles, and moved countless Xiamen people to tears. Many know that it took nine years to apply for World Heritage status, but few know that the Integrated Coastal Management (ICM) in Xiamen has gone through twenty years of hardship, a much longer and tougher process.

After the period of reform and opening, the Xiamen Special Economic Zone was established and the seas awaited planning. For all that time, the ocean has been the food basket for Xiamen's people. In the 1980s when aquaculture in Xiamen was at its peak, industrious fishermen contributed to the early economic development of the economic zone. However, good times did not last long. Excessive exploitation of sea resources for aquaculture and construction without adequate planning made the Xiamen Seas turbid, "Stinking puddles". Yundang Lake was a typical case. Therefore, since the mid-1990s, Xiamen has begun to implement ICM with the help of Partnerships in Environmental Management for the Seas of East Asia (PEMSEA) for marine pollution prevention and management projects. After twenty years of tenacious efforts, Xiamen gradually worked out a comprehensive coastal management framework tailored to the city's local conditions. The practice became an international benchmark. Chua Thia-Eng, "Father of Integrated Coastal Management", said: "The creation of the Xiamen model not only benefits future generations, but also provides a valuable reference for other countries and regions. This is an defining step in the history of ICM!"

※ A distinctive coordination mechanism has been established.

What is the core of such a grand proposition as ICM? For people working in this field, the answer, after twenty years' experience, is coordination.

In 1996, the overlapping of responsibilities among government agencies urgently needed resolution. In order to deal with the functional overlap and conflict among marine departments and law enforcement agencies, and to achieve marine and land integrated management, the Xiamen municipal government established an integrated coastal management and coordination mechanism. First, the government set up a Leading Group for Integrated Coastal Management with the mayor as the group leader and main leaders of marine departments as group members. The group was responsible for coordinating and dealing with major issues in coastal management. Second, the government established the Marine Management Office to organize the regular meetings of leaders from different marine departments and coordinate between marine departments so as to facilitate the resolution of hot and difficult issues in coastal management.

Two measures were taken to tackle the problem. The municipal government established the Xiamen Marine Management Office, a secondary institution under the Municipal General Office. Five years later, after coastal management had developed to a certain extent, the municipal government established Xiamen Municipal Bureau of Oceans and Fisheries, and the Marine Management Office was transformed into a coordinating institution. The Integrated Coastal Management was divided into two parts—an administration system and a coordination mechanism. The mechanism not only took charge of the industrial management, but also coordination on behalf of the government. So the Marine Management Office was kept under the Ocean Administration. The Office was headed by the Deputy Secretary General in charge of coastal management of the General Office of the government while its Executive Vice Director was the head of Xiamen Municipal Bureau of Oceans and Fisheries. Despite many rounds of reform, this mechanism is still in use today.

The Marine Management Office is an overarching unit that guarantees cooperation without limiting the functions or enthusiasm of other departments and units. With the Office promoting

▶ On November 14, 2015, workers were dredging the 18th flood ditch. It is said that in order to stop significant amount of silt from entering the Lake, dredging of silted basin and flood ditches in the Yundang Lake has been normalized and is carried out every year. (photo / Wang Xieyun)

active cooperation between departments, results can be more easily achieved. The merge of the Marine Management Office and the Aquatic Products Bureau laid the foundation for the development of Xiamen Municipal Bureau of Oceans and Fisheries.

Egoism is a fundamental part of human nature and cannot be eliminated. However, the coordinating mechanism of Integrated Coastal Management in Xiamen is not simply about shouting slogans, but about involving all people in the process by helping them realize that they could benefit from the ICM. This way, they will be more willing to cooperate. This reflects the collective wisdom of officials of the Xiamen municipal government in the process of coordination.

Looking at coastal management of Xiamen, we find that "integration" is the core, including the integration between government departments and the integration of technology and management. More importantly, all scientific research puts the management department first, and serves the administration. This has become a successful aspect of Xiamen's Integrated Coastal Management.

※ Local marine legislation has been established.

Nothing can be accomplished without norms and standards. Before marine laws were established systematically, Xiamen, as the first city to carry out Integrated Coastal Management, had to confront challenges with no guiding legal principles on several occasions. However, it did not hinder the coastal management, for Xiamen had explored a series of local marine regulations, and thus won precious time for Integrated Coastal Management.

In 1994 when Xiamen gained local legislative authority, the Municipal People's Congress and Municipal Government formulated more than ten marine laws and regulations such as

the *Provisions on the Use of Sea Areas of Xiamen, Provisions on Marine Environmental Protection of Xiamen, Provisions on the Protection of Chinese White Dolphins of Xiamen and Management Regulations on the Protection and Use of Uninhabited Islands of Xiamen*. All these form a complete local regulatory system under the national legal framework. Marine management has been on track legally since then. At the same time, the government formulated a series of marine projects, such as Xiamen Marine Functional Zoning, Marine Economic Development Plan of Xiamen and Xiamen Coastal Utilization Plan, forming a complete system of marine planning and management. This has laid a solid legal and administrative basis for the scientific and standard Integrated Coastal Management.

"Why do departments execute while the masses follow? What is the legal basis?" Wang Chunsheng, Director General of the Ocean and Fishery Administration of Xiamen, said frankly in an interview that, compared with seeking other technological and policy goals, marine legislation is more difficult and requires a combination of legal principles and attention to the reality of the water. The pressures, conflicts and obstruction met during this process were unimaginable except to those who experienced them.

In the preliminary stage of the legislation, the draft by the Marine Management Office was modified according to suggestions solicited internally regarding problems met in practical work and then handed off to the Legislation Bureau to be reviewed by legal experts. The introduction of a regulation usually underwent dozens of reviews and modifications.

Besides leading the country in marine legislation, Xiamen's ICM provides a practical model for national marine legislation to some extent. In the 1990s, Provisions on the Use of Sea Areas of Xiamen was issued and implemented. In 2001, the *Law of the People's Republic of China on the Administration of Sea Areas* was introduced. The State Oceanic Administration wrote to express gratitude to the Xiamen municipal government for its contributions. Several rules and regulations from *Provisions on Marine Environmental Protection of Xiamen* also have great significance.

Marine legislation of Xiamen must also keep pace with the times. It has improved with the passage of time. Sand excavation is an example. There were once several sand excavation sites, but after the implementation of Marine Functional Zoning, sand excavation was forbidden by law. For example, the Xiang'an Airport while it was under construction, needed sand. Xiamen applied for the designation of a zone for sand excavation on the shoal of the Taiwan Strait.

Not long after the Marine Management Office was established, a paramilitary team in charge

▲ Award for Outstanding Awareness Campaign of the Law

of marine supervision was formed to facilitate law enforcement at sea. This was a new law enforcement team created by Xiamen. The team uniform is like that of other city management officers, but the "Xiamen Marine Supervision Squadron" is named after the marine department. The brigade has its own certificate of administration and armband designed by the members themselves.

In order to solve the lack of coordination and administrative overlap at sea, Xiamen established a joint marine law enforcement mechanism. An "integrated marine administration law enforcement detachment" was formed to organize marine joint law enforcement activities. At the same time, the government explored ways to promote regional cooperation, resulting in the the establishment of the Xiamen-Zhangzhou-Quanzhou marine law enforcement city alliance, which aimed to coordinate the marine enforcement actions of three cities and solve coastal management problems across administrative areas.

Laws greatly facilitate administrative efforts. These laws must be observed and strictly enforced, and lawbreakers must be prosecuted. The policy instituted in the Third Plenary Session of the 11th Central Committee of the Chinese Communist Party was well implemented in the Integrated Coastal Management of Xiamen over the past two decades.

※ **Technical support has been secured.**

In the Integrated Coastal Management of Xiamen, members of the Xiamen Marine Expert Panel were not only specialists and scholars but also administrative staff and advocates. As a think tank, they provided strong support for the front-line staff. The elderly experts insisted on going to sea to work with others despite their old age. With their collective wisdom and hard work, they ensured the success of the coastal management of Xiamen, thus winning people's respect.

The accumulation of new managerial experience needs the support of new technology,

▶ The comprehensive management of small basins in Xiamen was progressing smoothly, with frequent good news and great changes in streams and rivers. The photo shows comprehensive management of water conservancy projects in the lower reach of Jiuxi River in Xiang'an. On July 28, 2016. (photo / Zhang Qihui)

and this requires constant communication between government and research institutes. Their opinions and demands can then be studied and evaluated. As the birthplace of oceanography in China, Xiamen is the center of marine research in Southern China. Xiamen is home to many prominent marine scientists and marine research institutions such as Xiamen University, the Third Institute of Oceanography of the State Oceanic Administration, Jimei University and Fujian Institute of Oceanography. The Xiamen municipal government set up the Marine Expert Panel to provide consultation and corroboration for coastal planning, development and ecological protection. They provided the scientific basis for decision-making of the municipal government and relative administrative departments. At the same time, our marine experts regularly use marine management mechanisms, geographic information systems and technology to carry out research projects on coastal management and protection. The Integrated Coastal Management was greatly improved as a result.

Before this, the Xiamen municipal government had set up several expert groups for land management. For example, there was an expert group in the department in charge of urban construction, and an expert group in the department in charge of science and technology. At that time, Hong Huasheng, professor at Xiamen University, and other experts, believed that forming too many expert groups would be harmful to coastal management efforts, and that only one expert panel was needed. The Expert Panel would not be necessarily large, but absolutely authoritative, with research efforts focused on the practical implementation of coastal management. Such an expert panel was more conducive to coastal management. The Marine Expert Panel was formed in response to these conditions.

In the preliminary stages of Integrated Coastal Management, the Marine Expert Panel members were mainly technological officials or NTOs. After the retirement of the first group of experts came the second stage. After the establishment of the Ocean Administration, the Expert Panel was mainly formed of scientific researchers to make sure that the group could execute its functions independently without any government intervention.

Integrated Coastal Management involves many disciplines. Every expert has his own expertise, so sometimes the Expert Panel will also have conflicting opinions. Generally speaking, experts see problems from a scientific perspective. Just like Ruan Wuqi, former director of the Fujian Marine Institute and the former leader of the Xiamen Marine Expert Panel, said, "Everyone in marine research needs to know their specialty and understand the relevant professional knowledge. " But sometimes, the suggestions made by the Expert Panel for development conflicted with those of the municipal government. The Integrated Coastal Management Leading Group must then evaluate the problem from an overall perspective.

It is no exaggeration to say that the effective implementation of Integrated Coastal Management in Xiamen is inseparable from the active participation of the Marine Expert Panel and the constant scientific and technological support it provides. The Expert Panel holds a meeting with leaders of the municipal government every two months or so. At the meeting, experts give feedback to the leaders on hot-button marine issues, and the leaders also express their views on marine issues and on how experts should give them technological support. The establishment of the Marine Expert Panel and the Marine Management Office truly realized the scientific coastal management in Xiamen and achieved notable results.

During the past few decades, Xiamen has completed many projects, from the Marine Functional Zoning to the information management system, from the comprehensive rehabilitation of Yundang lake to the planning and development of Wuyuan Bay, from the opening of the Gaoji Causeway to the development of Maluan Bay, from the establishment of the Marine Environment Monitoring Network to the setting of coastlines. All municipal marine projects are the results of the wisdom and sweat of the Xiamen Marine Expert Panel. Marine experts have provided a solid scientific and technological basis for the coastal management of Xiamen.

※ Marine economy has been developed and "marine ecological civilization" constructed

As the saying goes: "Those living by the mountains live off the mountains, and those living by the sea live off the sea", the key to the development of the economy of a city lies in the characteristic features of its economy adapted to its own conditions. The distinctive feature of an area fosters economic competitiveness and competitiveness brings strength. Xiamen's greatest advantage is the sea. Only by vigorously developing the marine economy can we develop the feature of Xiamen's economy. In the past 20 years, in the process of integrated coastal management in Xiamen, we have always sought the development of an "economic and ecological civilization".

In the past, people did not pay much attention to marine industry. There were only two main marine industrial models: aquaculture and shipping. Now, when the people of Xiamen go abroad, they see the beautiful sea in foreign countries and numerous marine industries. So, Xiamen decide to develop a cruise, yacht and sailboat economy first in Wuyuan Bay and marine biopharmaceutical industry from scratch to promote the development of emerging marine industries.

At the preliminary stage of Integrated Coastal Management, marine experts clarified the various responsibilities of the marine agencies with regards to the marine economy. Then the research group assessed the overall marine economic condition of the entire city and linked all kinds functions of the marine economy together and then put forward a targeted marine development program, which points out the goals for next year. The experts serve various departments and provide direction for development programs, whereas the industrial output of the programs belongs to the departments. This win-win model is an important part of Xiamen's initiative.

Before 2001, Xiamen proposed building a "bay city". After more than 20 years of economic development, the Xiamen Special Economic Zone ranked fifth in the country in terms of overall competitiveness, with many economic indicators at the forefront in the country. However, due to the restricted urban and economic output, as Xiamen is located on a small and isolated island, development has been limited. Therefore, it has become necessary to expand the space of the city and the development area through a transition from an "island city" to a "bay city". Only in this way can Xiamen improve its overall competitiveness, and sustain economic and social development. At the same time, after 20 years of construction, Xiamen must now develop beyond the main island due to limited development space and adjust its economic structure. It should also gradually move the manufacturing sector off of the island, focus instead on the development of modern services industry, and establish a free trade area on the island.

In recent years, the benefits brought by the Integrated Coastal Management of Xiamen are becoming more and more pronounced. The marine economy has maintained strong development momentum and has cultivated hundreds of billions of industrial chains. In 2015, the GDP of the city was 49.821 billion yuan, up by 13.9% over the previous five years, and the value added was 122 million yuan per square kilometer of sea area, 20 times the average level of the whole province and 80 times the national average. Xiamen is approved as a pilot base for national marine high-tech industry and a national demonstration site for marine science and technology.

Pan Shijian, former Deputy Mayor of Xiamen, once said that a unique feature of Xiamen is that the city focuses on harmony between man and nature by essentially promoting marine ecological restoration while also developing the economy.

While vigorously promoting traditional marine economy, such as shipping, coastal industries, ship-building, port logistics, and coastal tourism, Xiamen vigorously promotes emerging marine industries, such as "three ships" (cruise, yachts, and sailing) economy. (photo / Wang Huoyan)

Since the development of Integrated Coastal Management, more than 20 square kilometers of farmland were returned to the sea, there was about a 70 million increase in tidal prism, about 167 million cubic meters of accumulated dredging and about 11.2 square kilometers of land reclamation. All of these provide the basis for development of a marine "ecological civilization". Additionally, coastal beaches, mangrove forests and 30 km of coastline have been restored and about one million square meters of artificial beach constructed. In 2011, the Xiamen National Ocean Park became one of the first National Ocean Parks in the country. In 2013, after approval from the State Oceanic Administration, Xiamen became one of the first demonstration areas for the concept of building a "marine ecological civilization", and for national marine economic innovation and development. It also became the National Blue Bay demonstration city. Xiamen has ranked first for three consecutive years in the assessment of the quality of aquatic nature reserves by the Ministry of Agriculture, and ranked first for five consecutive years in the National Science Popularization Month on Aquatic Wildlife Protection. The "China Club Sailing Cup", "Strait Cup" and "College Students Sailing Cup" and other marine competitions demonstrate the prosperity and development of Xiamen's marine culture. The sailboat "Xiamen" completed the first circuit around the world of its kind by a Chinese sailing ship, carrying forward the boldness and strength of Xiamen.

At the same time, Xiamen also achieved fruitful results in foreign exchanges and cooperation. The China-ASEAN Maritime Cooperation Center, built by the State Oceanic Administration and the Fujian provincial government, was completed in Xiamen. PEMSEA shared the successful experience and achievements of the Integrated Coastal Management in the Xiamen demonstration area with other marine cities in East Asia. In particular, during the Xiamen

International Ocean Week, which has been successfully held for ten years in a row, activities such as the International Maritime Forum, marine exhibitions, business fairs and marine cultural displays were organized, further promoting Xiamen's popularity and influence in the world.

※ Public participation has been sought to secure public support

Integrated Coastal Management is an unending endeavor. Without support from the public and administrative coordination, no matter how sound the legal and regulatory system, and no matter how solid the scientific and technological support, they are nothing but castles in the sky. Fortunately, during the 20 years of coastal management of Xiamen, public participation has never been ignored.

Safeguarding the marine environment requires public involvement and support. The Xiamen Municipal Party Committee and the municipal government strengthened the publicity of marine laws and regulations by making full use of television, newspapers, radio and other news media to improve public awareness of marine environmental protection efforts and increase enthusiasm. Marine environmental protection activities were carried out on the World Wetlands Day, the World Ocean Day and the International Ocean Week in response to people's concerns and questions. In the construction of Yanwu Bridge, the government attentively listened to public opinion to make sure that the final bridge will not obstruct the view of Gulangyu, which won great praise from the masses.

In the Integrated Coastal Management of Xiamen, one of the most important aspects of public participation has been training. At that time, the State Oceanic Administration, the Xiamen municipal government and Xiamen University jointly built the Coastal Zone Sustainable Development Training Center, which was the predecessor to the Xiamen University School of Marine Affairs. The original training sessions were for officials from parallel and demonstration areas in the East Asian Seas, and all the member staff in the Administration must be trained regularly. The Fujian Institute of Oceanography is now working on training programs, mainly targeted at the third world countries.

In recent years, Xiamen and the State Oceanic Administration jointly built a national marine education network, with educational outreach based not only on the city's primary schools, middle schools and universities, but also on science and technology museums and other initiatives. The municipal government has also set up an education base for raising marine consciousness. Xiamen boasts the most marine education bases in the country. At the same time, scientific research ships and the Underwater World are also used by Xiamen Municipal Bureau of Oceans and Fisheries to publicize marine knowledge. For example, at Underwater World, after children watch dolphin performances, they are then given pamphlets showing them how to protect dolphin species, habitats, and the food chain. The Municipal Bureau of Oceans and Fisheries is also working with Tianxin Island Primary School in Haicang to compile marine scientific knowledge into textbooks. A special course has also been set up to popularize marine knowledge.

With the development of new media, new methods are now available to educate the public about the marine environment, management efforts and laws. In addition to cartoons, Xiamen Municipal Bureau of Oceans and Fisheries also spread information through WeChat, for example, by organizing a marine quiz show, where citizens with top scores were invited to visit the Chinese White Dolphin Rescue Base. The Municipal Bureau of Oceans and Fisheries is said to consider turning this kind of activity a regular program by introducing a credit system.

In recent years, Xiamen Municipal Bureau of Oceans and Fisheries's efforts to promulgate and publicize laws have been widely recognized by the masses and superior departments. In June 2016, Xiamen Municipal Bureau of Oceans and Fisheries was presented the "Award for Outstanding Group in Raising Public Legal Awareness" during the sixth five-year law publicity period from 2011 to 2015 by the Publicity Department of the CPC Central Committee, the Ministry of Justice and the National Law Publicity Office, the only coastal department to win such an award.

Twenty years of tenuous work by Xiamen coastal personnel can be condensed into five key pieces of experience, leaving a lasting imprint in the world history of coastal management. We have every reason to believe that in the coming 20 years, integrated coastal management will be carried to its highest pitch in Xiamen!

Summary of Scientific Research Programs During 9th Five-Year Plan Period to 12th Five-Year Plan Period of Marine and Island Management Office, Xiamen Bureau of Oceans and Fisheries

1. The 9th Five-Year Plan Period

(1) Provisions on the Use of Sea Areas of Xiamen

It is the first local marine regulation in China. This document has established three fundamental systems of marine functional zoning, coastal jurisdiction, and compensated use of coastal zone, which provides basis for law-based coastal management in Xiamen and its legal system in China.

(2) Xiamen Marine Functional Zoning

It is the first local marine functional zoning plan. This document specifies predominant functions, complementary functions, and restrictive functions of integrated coastal functional zones in Xiamen, and clarifies the legal status and force of Marine Functional Zoning. It is established as the scientific criteria of coastal utilization and management and provides experience for the establishment of marine functional zoning in China.

(3) Xiamen Management Measures for Coastal Utilization Fees Collection

It is the first supporting system for compensated use of coastal zone in China. These measures specify the concrete criteria for fees collection of every coastal zone for various purposes and provides operational basis for the compensated use system.

(4) Xiamen Marine Economy Development Plan

This plan integrates port, shipment, tourism, offshore industry, marine technology, and environment and integrated management in a bid to safeguard the comprehensive and coordinated development of marine industry.

(5) Xiamen Coastal Functional Geographic Information System

Based on the implementation demand of coastal hallmarks, coastal utilization certificate handling, counseling, technical support, marine environment monitoring and observation, database construction and macro scale coastal zoning improvement, GIS furnished with domestic leading caliber, high operability, and multiple interfaces has been developed, thus kicking off the stage of "digital seas".

2. The 10th Five-Year Plan Period

(1) Xiamen Uninhabited Islands Protection, Utilization and Management Measures

It is the first local regulation concerning uninhabited islands protection, utilization and management. Its launch has initially set up the management system of uninhabited island in Xiamen, and lays the groundwork for rule-basd and law-based management of uninhabited islands.

(2) Xiamen Coastal Mudflat Governance and Shelter Construction Plan

This plan has compiled coastal mudflat plan and shelter-belt construction plan, enhanced coastal shelter-belt construction and improvement, and arranged development, protection and management of coastal mudflat in Xiamen, based on research into the status quo.

(3) Xiamen Western Sea Areas Integrated Governance and Social and Economic Benefits Analysis

A framework for integrated governance and social and economic analysis is established to fully reveal information on western coast integrated governance effects. This provides basis for decision-making and integrated management.

(4) Maluan Bay Governance and Related Hydrological Dynamics Mathematical and Physical Model Research Report

This research delves into the practice of moderate governance measures to enhance tidal influx in western sea area so as to further improve water quality in western sea area, Maluan Bay, and Xinglin Bay, by boosting hydrological dynamics, diminishing sediment back-siltation in shipping channels, enlarging environmental capacity, and lifting water body self-cleaning capacity. While meeting the need of local planning on water body landscape and ecological environment, the research strives to realize the reasonable utilization of coastal areas and land resources, and provide scientific basis and technical support for western coast integrated management and development.

(5) Xiamen Uninhabited Islands Protection and Utilization Plan

It is the first local uninhabited islands protection and utilization plan. This plan has raised concrete requirements for island protection and utilization based on uninhabited island positioning and functions, which has provided foundation for detailed planning.

(6) Study on Xiamen Integrated Coastal Management Project

This document studies the experience of Xiamen Integrated Coastal Management (ICM), analyzes the environmental and resource bottlenecks from 3 aspects (pollutants, ecological system, and coastal land use), generalizes guidelines for ICM, puts forward long term and near term objectives, clarifies strategic management direction based on regional marine ecological system, and formulates measures and action for ICM.

(7) Xiamen Uninhabited Islands Ecological Rehabilitation and Landscape Construction Plan in Western Sea Areas

Studying and corroborating the feasibility of ecological rehabilitation and landscape construction, the plan helps achieve scientific protection and application of Xiamen uninhabited island resources by offering technical support and economic measures.

(8) Xiamen Coastal 1:5000 Water Depth Measurement

By conducting water depth measurement of Tong'an Bay, the Western sea areas, the Eastern sea areas, and Dadeng coastal areas of Xiamen, reliable and accurate macro scale water depth basic surveying and mapping materials can be obtained, which provide fundamental materials for scientific management and reasonable development of coastal area in Xiamen.

(9) Xiamen Coastal Management Geographic Information System

Upgrading Xiamen coastal functional GIS, this system effectively integrates GIS, GPS and RSG, introduces automatic process of approval procedures, updates fundamental marine data, and realizes new electrical version of functional zoning. The updated version continues to take the lead across the nation.

(10) Xiamen Coastal Zone Fees Collection Management Measures Revision Study

This study targets the status quo of fees collection in coastal areas of Xiamen, raises the utilization fees criteria which have basically covered all types of coastal zone utilization, takes the national lead in proposing fees collection for sea utilization projects like pollutant discharge, and formulates proposals for Xiamen coastal zone utilization fees collection and management.

3. The 11th Five-Year Plan

(1) Study on Coastal Utilization Right Tender and Auction Management Measures

This study analyzes the status quo of natural conditions and management operations of seas and islands around Xiamen, especially the present management of compensated utilization. It also explores the necessity and possibility of coastal utilization right tender and auction, legal basis included; explores procedures and plans, tender procedures and plans included, like tender formulation measures, contents, release, qualification verification, assessment methods, composition of tender assessment committee, and finalizing of tender candidate; explores procedures and plans of coastal utilization right auction; and delves into the linking issues between coastal utilization right tender and application, and the following implementation of law on land management in wake of obtaining coastal utilization right.

(2) Xiamen Uninhabited Islands Protection and Utilization Regulatory Detailed Plan

This plan has selected the area of over 2000 kilo square meters with certain development potential in uninhabited islands like Baozhuyu Island, Datuyu Island, Xiaotuyu Island, Houyu Island, Tuyu Island, Crocodile Island, Tuyu Island, and Dalipu Island into the planning, explores reasonable model of uninhabited islands protection and protective development and application, conducts land use planning, landscape planning, and corresponding municipal, and transportation supporting planning, puts forward corresponding environmental protection and ecological rehabilitation measures, formulates control index and requirements, and provides technical support for tender and auction.

(3) AGEI Historical Evolution of Marine Ecology in Xiamen West Sea Area and Database Construction Project

This project collects and integrates the archives of the geological survey of western sea area and the special marine survey in recent years in a comprehensive manner, conducts supplementary surveys on areas with little or no material, grasps background materials of this area (coastal belt included), and establishes a well-developed and effective marine geological environmental database which falls under the ICM implemented by the government.

(4) Xiamen Semi-closed Bay (West Sea Areas, Tong'an Bay) Aggregate Seal Area Control Study

This study analyzes influences of different coastline forms of the two bays on port shipment, environment and key ecological environments, combine effects of social, economic, environmental, and resources benefits, proposes reasonable area and scale of controlling western sea area and Tong'an Bay utilization from the angle of major control factors finalized by dominant functions of the bay, and provides scientific basis for coastal utilization and management.

(5) Gulangyu Island East Coast Shoal Governance and Treatment Plan Study

This study conducts survey on the beach landform status quo, completes 1:1000 water depth topographic survey within the studied area, rounds out geological survey, conducts artificial beach sand filling profile designing, computing, sand filling capacity computing, and ancillary works designing study, executes tidal current field numerical modeling computing targeted at sand filling experimental plan of artificial beach project, and accomplishes sand filling experimental project planning design.

(6) Xiamen Dalipu Island Neighboring Coast Regulatory Detailed Plan

This plan has conducted detailed planning and designing of the 310.34 square kilometers coastal area of Dalipu Islet and neighboring regions, proposed regulatory indicators of coastal utilization, and provided planning basis for market-based distribution of coastal utilization right.

(7) Xiamen Coastal Utilization Plan

On the basis of Xiamen Marine Functional Zoning, this plan has executed analysis on resource conditions, development potential, development and utilization status quo, existing issues, future economic development, and utilization requirements of different coastal areas, proposed demarcation suggestions on coastal utilization types, and finalized indicators, utilization intensity control research, coastal planning and layout, utilization scale, and management requirement.

(8) Xiamen Coastal Mud Flat Dredging Governance and Treatment Plan

This design proposes coastal mud flat dredging governance and treatment plan from aspects like increasing the tidal influx of Xiamen coast, enhancing the hydrodynamic conditions, maintaining and improving the water depth of the waterway in Xiamen port area, ensuring the sustainable development of the shipping and coastal tourism in Xiamen, improving the cultural landscape and the marine ecological environment, realizing the rational utilization of marine and land resources in the region, elevating city image, combining the elimination of mud flats, and reducing marine pollution.

(9) Supplementary Coastline Measurement

It has provided fundamental basis for the release of the statutory coastal line by conducting supplementary measurement of 1:5000 coastline in Xiamen.

(10) Xiamen Southeast Coastal Sand Extraction Zone Selection and Demarcation Study

This study has sought appropriate location for sand excavation, and provided supporting materials for grappling with huge demand and haphazard extraction in Xiamen by adopting blocking and dredging.

(11) East Coast Surrounding Region Coast Utilization and Coastal Belt Detailed Plan

This plan has conducted integrated analysis of east coast surrounding region and coastal belt from the perspective of determining the land according to the sea, and provided planning basis for integrated governance and treatment.

(12) Xiamen Island East Beach and Sand Source System Stability and Maintenance Measures Study

This study targets at the problems existing in the east beach of Xiamen Island, studies beach stability in this region, identifies different functional positioning of Xiamen Island east beach and proposes maintenance measures by consecutive seasonal monitoring and dynamic comparison of fixed profile, and surface sediment grain size feature analysis.

(13) East Coast Surrounding Region Integrated Governance and Treatment Project Intermediary Assessment Study

By tracking and assessing the implementation of the governance objectives in water quality, landscape, biomass, and hydrodynamics. this study objectively analyzes the effects of the integrated governance in east coast surrounding region.

(14) Xiamen Changwei Reef–Wutong Beach Artificial Beach Feasibility Analysis Study

This study has set up wind measurement station in Wutong, examined the effects of waves and sand blown by the wind on artificial beaches, analyzed the feasibility of artificial beaches by carrying out numerical modeling of Xiamen east coast, which has provided foundation for the project implementation.

(15) Xiamen Coast 3–Dimensional Model Study

This study has set up an informative, easy-to-use, interactive high performance and highly scalable 3-Dimensional GIS platform which, on the basis of remote sensing data, demonstrates operational data of different coastal utilizations, integrating satellite remote sensing data, aviation data and other dynamic monitoring data, and provides instruments for managing different sea areas.

(16) Xiamen Island Southeast Beach Dynamic System and Beach Recirculating Breeding and Protection Study

This study has conducted survey and analysis of beach quality in this area via historical data collection, monitored hydrodynamic conditions influencing waves, tides, hydrology and sediment, fathomed sand body cause, distribution and hydrological conditions which control sand breeding, and formulated sand rehabilitation project plans in Shizhoutou-Exhibition Center Coast Segment, Huangcuo Beach Segment, Baicheng Beach Segment, and Gulangyu Islet-Gangzai rear Beach Segment, which has provided scientific basis for beach renovation and protection, as well as reasonable utilization.

(17) East and West Coast Entrance Section Regular Observation and Xiamen Bay Water Power Improvement by Coastal Governance Assessment

This assessment performs seasonable observation of underwater topography, tidal level, and sediment in Songyu Islet-Xiagu Section and Wutong Section, Gaoji Section, and Weitou Bay. It also explores the positive role coastal governance has played in boosting the hydrodynamic conditions of the seas near Xiamen.

4. The 12th Five-Year Plan Period

(1) Xiamen Uninhabited Islands Rehabilitation Geological Conditions Study

This study has examined geological conditions in selected Baozhu Islet, Xiaotuyu Islet, Rabbbit Islet, and Crocodile Islet, conducted preliminary survey on neighboring stratum topography and sea area, analyzed stratum, topography, and their physical dynamics index of different islands, described geological conditions for filling fields, assessed island regional stability, slope stability, feasibility, and construction conditions, and conducted relevant economic analyses. Besides, surface soft earth properties are studied, with better understanding of grain diameter and landfill approach suggestions.

(2) Xiamen Dadeng Coast and Coastline Utilization Plan

This plan has utilized and regulated the coast and coastline in a reasonable manner, improved ecological environment, and provided basis for the management and regulation of other coasts in Xiamen.

(3) Xiamen Sand Beach Rehabilitation Technical Guide

Based on the research and practices in beach rehabilitation of Xiamen, the guide has summed up the general principle, content, working procedures, method and technical requirements during rehabilitation and proposed concrete approaches and measures throughout the whole process of investigating social demand, rehabilitation location selection, background survey, program design, model-based prediction, construction, and monitoring. It provides technical guidance for sand rehabilitation in Xiamen and other provinces and cities.

(4) Xiamen Bay Integrated Development Concept Plan

This plan conducts reasonable functional zoning and positioning of sea areas and coastal zones within the bay, divides the main functional areas in offshore areas, delimits specific scales of sea area utilization, proposes relevant controlling indicators, and provides science-based reference or foundation for marine and fishery administrative departments on coastal development administration; it also develops oceans in an orderly manner, strictly controls the intensity and type of development, relieves the pressure of various development activities on marine resources and the ecological environment, and safeguards space for ecological security and of landscape protection; furthermore, it takes marine environmental quality as a controlling indicator, appropriately expands the marine ecological space, achieves the matching of coastal economic layout with marine resources and environmental bearing capacity, and then enhances the sustainable development of the ocean.

(5) Dadeng Coastal Zone Utilization Plan

The plan analyzes the scale time sequence of development from the perspective of capacity against the background where large-scale land reclamation has basically accomplished in the western sea area and the surrounding eastern sea area, and the focus is about to be shifted to Dadong coastal area.

(6) Xiamen Coast Underwater Tourism Resources Survey

This study has harnessed underwater photographic equipment to explore underwater tourism resources in Xiamen coastal areas, and provides reference for maritime tourism.

(7) Xiamen Coast Dredging Engineering–Technical Specification and Acceptance Criterion

This criterion has combined the reality that Xiamen has entered the large-scale coastal dredging phase, and has provided guidance for regulation of dredging in Xiamen by formulating technical specification and acceptance criteria.

(8) Xiamen Dadeng Coast Land Reclamation Aggregate Control Study

The study has conducted analysis on utilization demand in Dedeng coast future development among existing plans, conducted numerical modeling under different scale reclamation plans, conducted assessment of different plans from the angle of hydrological dynamics, environmental capacity, ecology, marine resources, and economic returns, calculated comprehensive evaluation index of land reclamation impact, conducted aggregate control analysis, and determined reasonable scales for land reclamation at Dadeng coast.

(9) Xiamen Coast Ownership Management System Study

This study analyzes existing problems in first-tier and second-tier markets in coastal market management, and proposes relevant policy suggestions.

(10) Xiamen Coastal Utilization Fees Collection Criteria Study

This study establishes coastal value evaluation methods and coastal transfer value-added estimation methods to calculate key coastal area space resources utilization fees collection criteria, and sets up coastal value-added technical specification and management system. Based on this, Xiamen coastal utilization fees collection administration provides experience and demonstration for national coastal value assessment and value-added calculation.

(11) Xiamen Coastline Protection Plan

To put *Provisions on Marine Environmental Protection of Xiamen* in place, this plan has combined land territory positioning and layout planning in Xiamen coastal zone, and analyzed coastline demand in Xiamen. By conducting relevant studies, surveys and planning in coastline status quo, it has determined coastline functions and forms, conducted reasonable distribution and layout planning, specified scope of protection, defined restricted and prohibited development programs, and proposed relevant protective measures.

(12) Gaoji Causeway Opening and Dedeng Coast Reclamation Impact on Xiamen Hydrological Dynamics Observation Study

This study, based on the East and West Coast Section Regular Observation in 2010, has optimized plans, conducted hydrological dynamics features observation of key characteristic tides, mooring ocean current, and 26-hour ocean current profile in the spring of 2012. The study masters the dynamic impact process by Gaoji Causeway opening and Dadeng reclamation projects on coastal hydrological environmental temporal and spatial transition at Xiamen Bay.

(13) Xiamen Uninhabited Islands Census Registration Survey

Through topographic mapping, field survey and census registration of uninhabited islands, this survey formulates pictorial materials, which describe such conditions on uninhabited islands as topographical terrain, buildings, development and utilization.

(14) Xiamen Islands Management GIS

This system has established a system for island information collection, demonstration, and management, by integrated existing information resources on neighboring islands, drawing on mature design concepts of relevant systems, and taking advantage of advanced development technologies. This system has realized island GIS development and elevated Xiamen island management service level.

(15) Xiamen Sand Stability Analysis and Rehabilitation Study

From 2008 on, this study has conducted long-term beach profile observation, underwater beach topographic survey, and sediment survey and analysis on the beach of Xiamen. Based on this, it has mastered beach dynamic changes and analyzed the mechanism impacting shoreline stability, erosion and deposition, thus providing scientific basis for the further protection and rational development.

(16) Xiamen Coastal Tourism Resort Special Plan

In light of flourishing aquatic sports in Xiamen, this plan has conducted unified planning for existing coastal tourism resources based on cultural and environmental characteristics and delimited various thematic activity areas by matching natural and environmental conditions with coastal sports tourism programs, which provides basis for the future location selection of aquatic sports.

(17) Key Uninhabited Islands Protection and Utilization Plan

In accordance with national island plan compilation requirements, the plan has selected Datuyu Island, Huoshaoyu Island, and Crocodile Island which can be properly utilized, to compile protection plan. The plan has been registered at and won approval from the municipal government, providing "approval basis" for the development of these islands.

(18) Xiamen Coast Cumulative Impact Evaluation

The evaluation has comprehensively analyzed established ocean projects and monitoring projects in the past 5 years and complemented the existing projects with a cumulative impact study. It covers a cumulative impact evaluation of coastline, topography, hydrological dynamics, water quality, and ecological environmental evolution in western sea area, Tong'an Bay, Xinglin Bay, Maluan Bay and Wuyuan Bay.

(19) Xiamen Uninhabited Islands Tourism Wharf Engineering Project Design

The plan is designed to complement the island tourism development philosophy by the municipal government, and carries out the wharf design of 8 uninhabited islands including Chicken Island, Dayu Island, Datuyu Island, Huoshaoyu Island, Baozhuyu Island, Dalipu Island, Crocodile Island, and Tuyu Island.

(20) Xiamen Uninhabited Islands Tourism Project Planning

Based on major uninhabited islands status quo and neighboring coastal conditions surveys and studies like Chickeneu Island, Dayu Island, Datuyu Island, Huoshaoyu Island, Baozhuyu Island, Dalipu Island, Crocodile Island, and Shangyu Island, this planning has proposed tourism projects suitable for development in Xiamen islands, determined their functional positioning, and identified proper development projects, providing reference for decision-making and boosting healthy island tourism development.

5. The 13th Five-Year Plan Period

(1) Study on Revision of Provisions on the Use of Sea Areas of Xiamen

Taking into account the realities of coastal management, this study aims to supplement such content as aggregate control of coastal utilization, coastal zone purchase and storage, and ownership registration, which provides legal basis for furthering coastal management in Xiamen, and reference for the revision of the *Law on Coastal Zone*.

(2) Study on Xiamen Uninhabited Islands Protection, Utilization and Management Measures Revision

Against the backdrop of the promulgation of the *Law on Islands*, this study aims to revise the articles on the *Measures* that are incongruous to this law. Articles concerning inhabited islands ecological system protection will be added to improve the content concerning uninhabited islands protection, and article concerning moderate utilization of uninhabited islands will be refined to provide a legal basis for developing uninhabited islands in a moderate manner in Xiamen.

(3) Study on Xiamen Marine Functional Zoning Revision

Considering that the current provincial and municipal Marine Functional Zoning will not fully mature until 2020, during the mid-to-late period of the 13th Five-Year Plan, study on municipal marine functional zoning revision will be kick-started alongside provincial revision, which includes assessing current status quo of zoning implementation, analyzing marine utilization demands, giving priority to study on ecological system functional zoning, in a bid to incorporate the latest results into provincial and municipal Marine Functional Zoning.

(4) Coastal Use Right Market Deployment Benchmark Price Study

As the market becomes the predominant way of distributing coastal use right, based on current fees collection criteria, benchmark price for operational marine utilization types will be assessed in a bid to safeguard the state's yield right on coastal resources and enhance distribution efficiency. This study will provide value basis for market deployment.

(5) Xiamen Coastal Zone Utilization Regulatory Plan

Based on Xiamen Marine Functional Zoning, it will cater to the market-based distribution of costal use right, delimit various marine functional areas, and clarify development intensity, put up mandatory indicators, suggestion index, and management requirements targeted at different coastal utilization types, and provide guidelines for coastal utilization right market transfer.

(6) Coastal Zone and Islands Purchase and Storage System Study

Based on the study of national, provincial, municipal laws and regulations as well as domestic and overseas purchase and storage practices, in accordance with the realities of Xiamen, this study aims to set up a coastal zone purchase and storage system, with well-developed policies, laws and regulations and clarified responsible institutions, functions, management systems and procedures, so as to provide a better solution for coastal zone purchase and storage.

(7) Xiamen Bay Coastal Zone Spatial Layout Plan

In light of the "Greater Bay" development strategy of Xiamen, this plan aims to determine the future layout of coastal space around Xiamen, Zhangzhou, Quanzhou, and Jinmen, which will pave the way for building modern bays with distinctive features in China's coastal areas.

图书在版编目(CIP)数据

踏浪飞歌：1996—2016厦门海岸带综合管理二十年关键人物口述实录/厦门市海洋与渔业局,台海杂志社编. —厦门:厦门大学出版社,2018.7
ISBN 978-7-5615-6926-9

Ⅰ. ①踏… Ⅱ. ①厦… ②台… Ⅲ. ①海岸带-综合管理-概况-厦门-1996—2016 Ⅳ. ①P748

中国版本图书馆CIP数据核字(2018)第074245号

出版发行	厦门大学出版社
社　　址	厦门市软件园二期望海路39号
邮政编码	361008
总 编 办	0592-2182177　0592-2181406(传真)
营销中心	0592-2184458　0592-2181365
网　　址	http://www.xmupress.com
邮　　箱	xmup@xmupress.com
印　　刷	福州报业鸿升印刷有限责任公司

开本　787 mm×1 092 mm　1/16
印张　36.75
字数　780千字
版次　2018年7月第1版
印次　2018年7月第1次印刷
定价　116.00元(中英文)

本书如有印装质量问题请直接寄承印厂调换

厦门大学出版社
微信二维码

厦门大学出版社
微博二维码